Corrosion Resistant Alloys for Oil and Gas Production: Guidance on General Requirements and Test Methods for H₂S Service

T0362179

European Federation of Corrosion Publications

NUMBER 17 Second Edition

A Working Party Report on

Corrosion Resistant Alloys for Oil and Gas Production: Guidance on General Requirements and Test Methods for H$_2$S Service

CRC Press
Taylor & Francis Group
Boca Raton London New York

CRC Press is an imprint of the
Taylor & Francis Group, an **informa** business

Published for the European Federation of Corrosion
by CRC Press on behalf of The Institute of Materials

First published 2002 by Maney Publishing Ltd

Published 2018 by CRC Press
711 Third Avenue, New York, NY 10017, USA
2 Park Square, Milton Park, Abingdon, Oxon OX14 4RN

CRC Press is an imprint of the Taylor & Francis Group, an informa business

Typeset in India by Emptek, Inc.

ISBN 13: 978-1-902653-55-6 (pbk)

Contents

European Federation of Corrosion Publications
Series Introduction

The EFC, incorporated in Belgium, was founded in 1955 with the purpose of promoting European co-operation in the fields of research into corrosion and corrosion prevention.

Membership is based upon participation by corrosion societies and committees in technical Working Parties. Member societies appoint delegates to Working Parties, whose membership is expanded by personal corresponding membership.

The activities of the Working Parties cover corrosion topics associated with inhibition, education, reinforcement in concrete, microbial effects, hot gases and combustion products, environment sensitive fracture, marine environments, surface science, physico-chemical methods of measurement, the nuclear industry, computer based information systems, corrosion in the oil and gas industry, and coatings. Working Parties on other topics are established as required.

The Working Parties function in various ways, e.g. by preparing reports, organising symposia, conducting intensive courses and producing instructional material, including films. The activities of the Working Parties are co-ordinated, through a Science and Technology Advisory Committee, by the Scientific Secretary.

The administration of the EFC is handled by three Secretariats: DECHEMA e.V. in Germany, the Société de Chimie Industrielle in France, and The Institute of Materials in the United Kingdom. These three Secretariats meet at the Board of Administrators of the EFC. There is an annual General Assembly at which delegates from all member societies meet to determine and approve EFC policy. News of EFC activities, forthcoming conferences, courses etc. is published in a range of accredited corrosion and certain other journals throughout Europe. More detailed descriptions of activities are given in a Newsletter prepared by the Scientific Secretary.

The output of the EFC takes various forms. Papers on particular topics, for example, reviews or results of experimental work, may be published in scientific and technical journals in one or more countries in Europe. Conference proceedings are often published by the organisation responsible for the conference.

In 1987 the, then, Institute of Metals was appointed as the official EFC publisher. Although the arrangement is non-exclusive and other routes for publication are still available, it is expected that the Working Parties of the EFC will use The Institute of Materials for publication of reports, proceedings etc. wherever possible.

The name of The Institute of Metals was changed to The Institute of Materials with effect from 1 January 1992. The series is now published by Maney Publishing on behalf of the Institute of Materials.

A. D. Mercer
EFC Series Editor, The Institute of Materials, London, UK

Mr. R. Mas
Fédération Européene de la Corrosion, Société de Chimie Industrielle,
28 rue Saint Dominique, F-75007 Paris, FRANCE

Professor Dr. G. Kreysa
Europäische Föderation Korrosion, DECHEMA e.V., Theodor-Heuss-Allee 25
D60486, Frankfurt, GERMANY

EARLIER VOLUMES IN THIS SERIES

Preface to the First Edition

This Working Party Report is the companion to EFC 16 *Guidelines on Materials Requirements for Carbon and Low Alloy Steels for H₂S-Containing Environments in Oil and Gas Production*. These reports have been produced by Work Groups in the Working Party on Corrosion in Oil and Gas Production since it was formed in 1992.

The driving force for the preparation of this report has been the long standing, unsatisfactory inconsistency in testing and qualifying corrosion resistant alloys (CRAs) for H₂S service. The primary problem was considered to be that there was no standard methodology for establishing the environmental cracking resistance of CRAs in H₂S service. Improving this situation became the CRA Work Group's initial aim.

The report therefore proposes test methods for assessing the environmental cracking resistance of CRAs. In order to document the basis for these proposals, extensive background information has been included along with further information related to the use of CRAs in oil and gas production.

However, to produce the report in a reasonable time it has been necessary to limit its scope. Thus, it has not been possible to include detailed test methods for pitting and crevice corrosion or develop guidance on the service limits of individual CRAs. These important tasks, among others, remain for consideration in future revisions or reports.

The CRA Work Group has been well supported by all sections of the industry that have an interest in the use of CRAs. As chairmen of the EFC Working Party and Work Group that have produced this report, we wish to thank all who have supported the work. This includes sponsorship of Work Group members by employers, provision of meeting facilities by host organisations and contributions from individuals. Unfortunately, the contributors are too numerous to name individually, though NiDI's essential organisational and secretarial sponsorship warrants particular mention. Working with the group, whose membership has been drawn from Europe and beyond, has been very rewarding for us. We hope readers find value in the product of this labour.

Liane Smith
Chairman
European Federation of Corrosion
Working Party on Corrosion in
Oil and Gas Production

Bill Nisbet
Chairman
Corrosion Resistant
Alloys Work Group
of the Working Party

Ed Wade
Past Chairman (1992-1996)
Corrosion Resistant
Alloys Work Group
of the Working Party

Preface to the Second Edition

The work group consider this revision of EFC 17 necessary to incorporate developments, in the testing of CRAs, made since publication of the first edition in 1995. In particular:

1. Experience with weldable super-martensitic stainless steels has identified that artificially buffered test solutions used for SSC testing require modification for these steels
2. More general improvements in the definition of test solutions have been made in co-operation with ISO/TC 67/WG 7 during the preparation of ISO 15156.

The above have been incorporated as changes to Section 8 and Appendix 4. Elsewhere, minor changes have been made to update and correct editorial errors and omissions in the original text.

In addition to the changes now made to this document, the reader's attention is drawn to:

1. Continuing development of ISO 15156 by ISO/TC67/WG 7.
2. Extensive proposed changes to the CRA content of NACE MR0175 and the intended carry-over of these changes into ISO 15156.
3. Incorporation of SCC testing requirements in the 1996 revision of NACE TM0177 and changes to test solutions now in preparation for the next edition. The latter are expected to be largely consistent with this document.
4. Publication of ISO 13680:2000 that defines 'technical delivery conditions' for CRA OCTG not previously covered by API 5CT (ISO 11960).

We wish to acknowledge the essential contributions of Work Group members to this revision and the following who have assisted with the editorial preparation of the revision: J.-L. Crolet, E. Wade, B. Kermani.

As foreseen in the first edition, developments in the testing of CRAs continue. Readers are encouraged to minitor these to ensure their use of best practice.

Phil Jackman
EFC Oil and Gas Chairman

Chris Fowler
Working Party Chairman

Mike Swidzinski
Working Party Chairman

1

Terminology and Definitions

All terminology in this document is advisory, terms such as: shall, should; require, requirement; etc., are not to be taken as mandatory when there are sound technical reasons for acting otherwise.

The document uses established oilfield terminology. Some key generic terms are explained in Section 1.1 below. Conventional definitions of other terms and abbreviations follow in Section 1.2.

1.1 Generic and Oilfield Terms

Oilfield and Oilfield Facilities

'Oilfield' and 'oilfield facilities' should be taken to include oil, condensate and gas producing facilities.

Carbon Steels

'Carbon steels' is used as a generic term to designate the full range of carbon, carbon-manganese and low alloy steels, used in the construction of conventional oilfield equipment.

Corrosion Resistant Alloys (CRAs)

CRAs are alloys that are intended to be resistant to general and localised corrosion in oilfield environments that are corrosive to carbon steels, as defined above.

Sweet Service / Sour Service

Conventionally these terms are used to distinguish between service in H_2S free (sweet) and H_2S containing (sour) fluids; the distinction is commonly made in accordance with the criteria of NACE MR0175. Because there is no satisfactory standard definition of these terms as they apply to the use of CRAs, the terms are not used in this document. Instead the term H_2S service is used to identify environments that contain measurable amounts of H_2S. H_2S is conventionally measured in gas streams. Measurement down to 0.5 ppm by volume is routinely available in accordance with ASTM D 4810.

Environmental Cracking, Sulphide Stress Cracking, Stress Corrosion Cracking

This document proposes test methods for assessing the resistance of CRAs to environmental cracking in H_2S service. Cracking may be caused by sulphide stress cracking (SSC) or stress corrosion cracking (SCC) which occur by different mechanisms as described in Appendix 1. As the objective of the tests is to identify conditions that cause cracking, irrespective of the mechanism, the tests are referred to throughout as SSC/SCC tests unless the immediate context requires that the cracking mechanisms be differentiated.

Titanium Alloys

The term titanium alloys includes the commercially pure grades.

Test Methods and Specimens

The terminology of ISO 7539-1 has been followed in describing SSC/SCC test methods and test specimens.

Yield Strength, Proof Strength, Elongation under Load

These terms cause confusion because they are defined and used differently in Europe, the USA, ISO and oilfield specifications. In this document, yield strength and yield stress have been used, according to the ASTM and oilfield conventions, to indicate the stress (factored for design purposes) at which significant plastic yielding occurs in a tensile test. This usage is an imprecise convenience as the strength cannot be quantified without specifying the extent and method of measurement of the plastic deformation. In the proposed SSC/SCC test procedures the recommended test stresses are defined rigorously according to ISO convention.

In this context it should be noted that for some commonly used CRAs, API standards specify measurement of yield at 0.5% extension under load (EUL). Examples are Grade L80 13Cr casing and tubing in API 5CT and the CRA line pipe grades in API 5LC. By contrast, the ISO draft specification for CRA casing and tubing (ISO/WD 13680-1) uses the 0.2% proof stress to define 'yield'.

Although the difference between these two definitions of proof stress is typically insignificant in carbon steels and 13Cr, the difference may be significant for alloys that have a high strain hardening rate. This includes many stainless steels and nickel base alloys to which this document applies.

1.2 Definitions and Abbreviations

CRA	Corrosion resistant alloy, see also Section 1.1 above.
Constant load, constant total strain	1SO 7539 terms for methods of loading SCC specimens. C-rings and 4 point bent beams are conventionally loaded in constant total strain.
Compliance	Elastic stiffness.
ε_{air}	Strain to failure in air in SSRT.
ε_n	Normalised strain to failure = $\varepsilon_s/\varepsilon_{air}$ in SSRT.
ε_s	Strain to failure in solution in SSRT.
EDTA	Ethylene diamine tetra-acetic acid.
HAZ	Heat-Affected Zone. That portion of the base metal that was not melted during brazing, cutting or welding but whose microstructure and properties were altered by the heat of these processes.
LTC	Low temperature creep.
OCTG	'Oil country tubular goods', i.e. oil well casing, tubing and drill pipe.
Partial Pressure	Ideally, in a mixture of gases, each component exerts the pressure it would exert if present alone at the same temperature in the total volume occupied by the mixture. The partial pressure of each component is equal to the total pressure multiplied by its mole or volume fraction in the mixture.
Plastic Deformation	Permanent deformation of a material caused by stressing beyond the limit of elasticity, i.e. the limit of proportionality of stress to strain.
PEEK	Polyetheretherketone.
ppm	Parts per million (i.e. $\times 10^{-6}$).
ppb	Parts per billion (i.e. $\times 10^{-9}$).

PRE	Pitting Resistance Equivalent. There are several variations of these indices which are usually based on observed resistance to pitting corrosion of CRAs in the presence of dissolved chlorides and oxygen, e.g. sea water. Though useful, these indices are not directly indicative of corrosion resistance in H_2S-containing oilfield environments. Originally they were evaluated as a function of %Cr and %Mo but now are commonly extended to include other elements. Examples of expressions in current use are:

$$PREN = \%Cr + 3.3\%Mo + 16\%N, \text{ and}$$

$$PRENW = \%Cr + 3.3\,(\%Mo + 0.5\%W) + 16\%\,N.$$

Proof stress	The stress at which the non-proportional elongation is equal to a specified percentage of the original gauge length. Denoted by R_{px} where X is the specified percentage and equivalent to the 'offset method' of yield strength determination in ASTM A370.
P_{H_2S}	Partial pressure of H_2S in a gas mixture.
PTFE	Polytetrafluoroethylene.
QA	Quality Assurance.
QC	Quality Control.
RA_{air}	Reduction of area of a tensile specimen in air in SSRT.
RA_n	Normalised Reduction of area of tensile specimens = RA_s/RA_{air} in SSRT.
RA_s	Reduction of area of a tensile specimen in a test solution in SSRT.
R_p	See 'proof stress' above.
SCC	Stress corrosion cracking, i.e. cracking caused by the combined action of corrosion and tensile stress. See 1.1. above.
SCE	Saturated Calomel Electrode

SEM   Scanning electron microscope.

SSC   Sulphide stress cracking, i.e. cracking under the combined action of corrosion and tensile stress in the presence of water and hydrogen sulphide. See 1.1. above.

SSRT   Slow strain rate test.

Sustained load   A method of loading specimens by use of a proof ring or similar device.

Yield Strength   See Section 1.1.

2

Standards Referred to in this Document

API 5CT	Specification for Casing and Tubing.
API 5LC	Specification for CRA Line pipe.
ASTM A 370	Mechanical Testing of Steel Products.
ASTM D 513	Standard Test Methods for Carbon Dioxide and Bicarbonate and Carbonate Ions in Water.
ASTM D 4810	Standard Test Method for Hydrogen Sulphide in Natural Gas Using Length-of-Stain Detector Tubes.
ASTM D 1193	Standard Specification for Reagent Water.
ASTM G 38	Standard Practice for Making and Using C-Ring Stress Corrosion Test Specimens.
ASTM G 39	Standard Practice for Preparation and Use of Bent-Beam Stress-Corrosion Test Specimens.
BS 6888	British Standard Methods for Calibration of Bonded Electrical Resistance Strain Gauges.
ISO 15156	Petroleum and natural gas industries - Materials for use in H_2S - containing environments in oil and gas production. Part 1 published 2001 parts 2 & 3 in preparation.
ISO 13680:2000	Petroleum and natural gas industries - Corrosion-resistant alloy seamless tubes for use as casing, tubing and coupling stock - technical delivery conditions.
ISO 11960:2001	Petroleum and natural gas industries - steel pipes for use as casing or tubing in wells.
DIN 50 905	Corrosion of Metals. Corrosion Investigations (in German).

ISO 6892   Metallic Materials- Tensile Testing.

ISO 7539:   Corrosion of metals and alloys - Stress corrosion testing.

   Part 1:   General guidance on testing procedures.

   Part 2:   Preparation and Use of Bent-Beam Specimens.

   Part 5:   Preparation and Use of C-Ring Specimens.

   Part 7:   Slow Strain Rate Testing.

NACE TM0177-96   Standard Test Method. Laboratory Testing of Metals for Resistance to Specific Forms of Environmental Cracking in H_2S Environments.

NACE MR0175-2001   Standard Materials Requirements. Sulphide Stress Corrosion Cracking Resistant Metallic Materials for Oilfield Service.

NORSOK   MP-DP-001 Design Principles - Materials Selection.

NACE TM0198-98   Slow Strain Rate Test Method for Screening Corrosion-Resistant Alloys (CRAs) for Stress Corrosion Cracking in Sour Oilfield Service

3

Introduction

Traditionally, 'carbon steels' have been the principal bulk materials used for construction of oil and gas field production facilities. Although these remain the materials of first choice on the grounds of cost and availability, the oil and gas industry is increasingly using corrosion resistant alloys (CRAs) for the construction of primary production equipment. This trend results from a combination of pressures which include: the production of more corrosive fluids, more hostile operating locations, a requirement for improved equipment reliability with associated, escalating, safety and environmental considerations. Simultaneously, there is remorseless pressure on the industry to reduce costs. Thus, increasingly the trend is to seek the lowest overall cost of equipment ownership, rather than just its lowest initial cost. Whilst CRAs may offer the lowest life cycle cost, they carry a heavy initial cost penalty, compared with, traditional 'carbon steel' equipment. Clearly, it is important to ensure that the most economic CRAs are specified whenever they are required.

It follows that laboratory test methods used to qualify alloys for service are required to be sufficiently discriminating to identify candidate materials correctly without incurring the cost penalties of over-specification or the performance shortfalls of under-specified alloys. This requires detailed understanding of the intended service conditions and careful definition of corrosion test requirements.

To date, there has been no generally accepted methodology for testing and selecting CRAs for oilfield service. The lack of a common methodology has been particularly troublesome (and costly) in the case of qualifying alloys for H_2S service. The practices established for qualifying carbon steels are generally not applicable to CRAs. In the absence of an established methodology, manufacturers, designers, test laboratories and operators have had different and conflicting opinions of the test requirements for CRAs. Inappropriate testing has resulted in conflicting data and suboptimal selections.

This document attempts to improve this unsatisfactory position by establishing a common understanding of the requirements for the testing and qualification of CRAs for use in oil and gas field production facilities. It provides a general summary of the issues involved in selecting and qualifying CRAs, and proposes test methods for evaluating the threshold conditions for environmental cracking (SSC/SCC) of CRAs in H2S service. Background information, important to the process of testing and selecting CRAs, is also provided.

4

Scope

This document addresses the selection and qualification of CRAs for use in oil and gas field production facilities that handle raw and partly processed reservoir fluids at, and below, reservoir temperatures. It does not address the requirements of downstream processing facilities such as refineries.

The document is concerned with the resistance of candidate alloys to corrosion and environmental cracking. Mechanical properties are outside the scope of the document.

The test procedures proposed for evaluating environmental cracking (by SSC/SCC), are applicable to stainless steels, nickel alloys, and titanium alloys. As currently proposed, these test procedures do not account for:

i.   Galvanic coupling (including cathodic protection) which can be detrimental to the cracking resistance of CRAs.

ii.   The presence of liquid mercury in produced fluids.

iii.   Complex circumstances in which one environment produces a susceptibility to cracking which subsequently occurs under different environmental conditions. For example, the possibility of hydrogen charging at elevated temperature, causing embrittlement after cooling, is not considered.

iv.   The effect of elemental sulphur on environmental cracking has not been addressed in the main document but Supplementary Appendix S1 prepared independently by the German Institute for Iron and Steel (VDEh), is included as a first step towards defining test procedures for fluids containing elemental sulphur.

The document is intended to provide a basis for future development of CRA testing and selection. It has been written to provide guidance and is not suitable for mandatory application. Engineering judgement and interpretation is required in the application of the document. In particular, the document is not intended for retrospective application to existing facilities that operate satisfactorily.

5

Objectives

The aims of this document are:

i. To establish a common understanding of hydrocarbon production environments, as a basis for specifying corrosion test environments for qualifying CRAs for oilfield service.

ii. To identify and summarise the general requirements for corrosion resistance of CRAs intended for oilfield service.

iii. To establish environmental cracking (SSC/SCC) test methods to qualify CRAs for oilfield H_2S service.

iv. To provide background information that is important to testing, qualifying, selecting and specifying the corrosion resistance of CRAs intended for oilfield service.

v. To provide a documented basis for the further development of corrosion test procedures for oilfield CRAs.

6

Overview of Requirements for Selection and Qualification of CRAs for Oil and Gas Field Use

6.1 Introduction

Conventionally, the use of CRAs is considered when the corrosivity of the produced fluids makes carbon steels uneconomic. It follows that the primary requirement for CRAs is adequate resistance to corrosion by the water phase of produced fluids. In addition, to have practical use, CRAs must be compatible with the other environments encountered in oil and gas field production facilities. This section provides an overview of these requirements.

6.2 Resistance to Corrosion by Produced Fluids

6.2.1 Carbon Steels

The water phase of raw, produced fluids, although free of dissolved oxygen, may be made corrosive to carbon steels by acidic components associated with the hydrocarbons. These are principally the acid gases CO_2 and H_2S, and in some cases organic acids. Acid corrosion of carbon steels caused by these components may take the form of general or localised corrosion (involving significant weight loss) or sulphide stress cracking. The corrosivity of fluids is determined by the acid components and the natural water composition, as described in Appendix 2. The corrosivity may be modified, intentionally or unintentionally, by the conditions of the primary production and processing facilities. Corrosivity to carbon steels, (and more generally to other alloys), may be:

- reduced by lower temperature[1], lower pressure, corrosion inhibitors, etc.; or:
- increased by high fluid velocities, oxygen ingress, microbiological contamination, etc.

6.2.2 CRAs

By definition, when used in appropriate environments, CRAs are either fully resistant to general corrosion (by virtue of a passive, protective surface oxide film); or their

---

[1] For carbon steels, the normal effect of temperature may be reversed by changes in the form and stability of corrosion product films.

corrosion rates are sufficiently low to be acceptable over the intended service life of the equipment. Also, to justify their cost, CRAs are required to be resistant to:
- localised corrosion in the form of pitting and crevice corrosion;
- environmental cracking (SSC/SCC).

For any particular duty, these minimum requirements must be met by candidate CRAs.

The principal environmental parameters that determine the corrosion performance of CRAs are:
- in situ pH (of the water phase);
- chloride concentration;
- H_2S concentration;
- temperature;
- the presence of elemental sulphur ($S°$);
- processing contaminants such as oxygen.

The pH is determined by the amounts of CO_2 and H_2S present and the water composition as detailed in Appendix 2.

6.2.3 Testing CRAs for General and Localised Corrosion

Establishing that an alloy has adequate resistance to general and localised corrosion is an important pre-requisite to performing the more complex environmental cracking tests that are proposed in this document. Alloys that have a marginal resistance to environmental cracking will be prone to crack after extended exposure in conditions that cause localised corrosion, as a result of active local corrosion and the consequential stress concentrations.

For CRAs, the worst conditions for general and localised corrosion are normally associated with the maximum service temperature. Tests to establish adequate resistance therefore require definition and reproduction of the environmental parameters at that temperature. The characteristics of CRAs make it difficult to specify practical short term test conditions, that are adequately conservative to represent long term service exposure, without their being unrealistically severe. Because the corrosion resistance of CRAs tends to exhibit pronounced thresholds to environmental parameters, escalation of one or more test parameters in an attempt to accelerate tests, may result in severe corrosion and unnecessarily exclude a potentially satisfactory alloy.

In addition to defining appropriate environmental test conditions, the following also require consideration to assess fully the serviceability of an alloy.

i. How are 'worst case' material compositions and microstructures to be accounted for in tests?

ii. What is the critical property to be tested? For CRAs subject to localised corrosion, is it:

- resistance to pitting of uncreviced surfaces;
- resistance to crevice corrosion?

iii. What should be the condition of the test surface?
- the 'ex mill' or 'as supplied' finish;
- a machined or otherwise modified surface;
- 'end grain';
- welded material?

iv. Should recovery from mechanical or chemical depassivation be included in the tests? (Wirelines and mineral acids are respectively examples of potential mechanical and chemical depassivators).

This document does not address the details of test procedures for determining the general and localised corrosion resistance of CRAs as a variety of test methods are established and described in literature. However, the recommendations of this document, concerning appropriate test environments for environmental cracking tests, are also applicable to the evaluation of general and localised corrosion resistance.

6.2.4 Testing CRAs for Resistance To Environmental Cracking

Generally, the same environmental parameters, as previously listed in 6.2.2., also determine the resistance of CRAs to environmental cracking. However, the magnitude and nature of the local tensile stress, and the material condition are additional parameters that control cracking. Alloys that are susceptible to SCC are most likely to crack at their highest exposure temperatures. The presence of H2S will, generally, increase their susceptibility to cracking. In contrast, the temperature of greatest susceptibility to SSC occurs at lower temperatures, typically around 25°C for carbon and low alloy steels. However, because cracking of CRAs may result from interaction of the different mechanisms, some alloys may exhibit their maximum susceptibility to cracking at intermediate temperatures. A fuller explanation is given in Appendix 1.

Test methods, proposed for determining SSC/SCC resistance, are addressed in detail in Sections 7 and 8.

6.3 Corrosion Resistance in Other Fluids

In addition to resisting corrosion by raw, produced fluids, candidate alloys are required to be resistant to other fluids to which they are exposed during production operations. These may be natural fluids (such as sea water), manufactured chemicals (such as brines and mineral acids), or combinations of both types. The majority of these fluids have a history of use with conventional equipment. In many cases their compatibility with CRAs is not well known and requires review when candidate alloys are being considered. Corrosion resistance needs to be assessed with respect to both metal loss and environmental cracking resistance.

Commonly used fluids include:

i.   Drilling muds, diesel oil, drilling base oil, etc.

ii.   Synthetic completion and kill fluids. (These are normally brines based on NaCl, KCl, $CaCl_2$, NaBr, $CaBr_2$, $ZnBr_2$, etc.; initially oxygenated by atmospheric exposure).

iii.   Sea water (normally oxygenated by atmospheric exposure).

iv.   Acids (acetic and citric acids, HCl, HCl/HF). (Note: H_2S may be evolved if acids contact sulphide scales).

v.   Scale dissolvers (acids, EDTA).

vi.   Asphaltene dissolvers (organic solvents).

vii.   Elemental sulphur dissolvers.

viii.   Inhibitors for corrosion, scale and wax control. (Note: squeeze inhibitors may occur at high concentrations.)

ix.   Methanol and glycols for control of hydrates.

In addition to these fluids that may be introduced into the production stream, CRAs may also be exposed to secondary environments such as those that occur in well annuli, heat exchangers, atmospheric exposure, etc. Although detailed consideration of the corrosion properties of these fluids and environments is beyond the scope of this document, it is worth noting the following.

i.   Exposure times are generally short for fluids introduced into production streams and, therefore, the duration of corrosion tests may approximate the anticipated service exposure. This means that accelerated tests are not generally required.

ii.   In production streams, fluids may be used in combination or sequentially and produced fluids may be present with significant concentrations of production chemicals (e.g. natural H_2S production may coincide with low pH acid returns).

iii.   Titanium alloys are subject to some particular limitations as summarised in Appendix 3.

The consequences of ignoring the potential incompatibility of CRAs with 'non-produced' fluids and secondary environments can be severe.

6.4 Qualification and Ranking of CRAs

The corrosion resistance of an alloy may be 'qualified' for a particular service either:

i.   by testing in an environment that simulates the specific intended service; or,

ii.   by acceptance of the results obtained from tests performed in an equivalent or more severe environment.

To minimise the requirement for application specific testing, there is a need for test data, obtained under appropriate 'standard' conditions, that permit use of alloys in commonly encountered service environments without further testing. Results from such tests are also useful as 'ranking' tests for identifying candidate alloys for testing in specific service environments.

To this end, 'reference' environments for SSC/SCC testing of CRAs are proposed in Appendix 4. These environments are, in principle, also appropriate for establishing the resistance of CRAs to general and localised corrosion.

6.5 Quality Assurance (QA), Quality Control (QC) Testing

In addition to the need to qualify alloys for their intended duty, there is a requirement to verify that the material supplied has the intended properties. It is important that QA/QC tests are specified to achieve this and not used as qualification tests. A clear distinction should be made between qualification tests and QA/QC tests.

Qualification testing should be complete before commercial orders are placed, and QA/QC tests used during manufacture only to confirm that the material is produced to the specified standard or requirements. The criteria for inclusion of corrosion tests for acceptance purposes should be:

i.   That they measure properties that are explicit requirements of the manufacturing specification.

ii.   That test failures can be explained and traced to controllable manufacturing parameters so that corrective action can be taken by the supplier.

iii.   That standard test procedures exist that are known to be reliably reproducible.

This approach is considered to separate correctly the responsibilities of the user and the manufacturer.

Requirements for QA/QC testing are outside the current scope of this document.

7

General Principles and Limitations of Proposed SSC/SCC Tests

7.1 Background

This section describes the principles of the tests that are considered appropriate for establishing the environmental limits for reliable use of CRAs in H_2S service. Test methods are proposed and their practical limits are identified. The core principle is that materials should be evaluated under the most severe environmental and mechanical conditions that are realistically anticipated for the intended service and not under standard 'worst case' conditions. The degree of conservatism involved in specifying the test conditions should reflect the consequences of service failures.

This approach allows alloys, that are not inherently immune to SSC/SCC, to be used reliably, with economic benefit, in a restricted range of mechanical and environmental conditions. The requirement is to define tests which identify these application limits without simply rejecting alloys that are susceptible to SSC/SCC under extreme test conditions.

7.2 Test Environments

For the reasons previously discussed, use of a single test environment to qualify CRAs for all H_2S service conditions is inappropriate as it will inevitably result in either uneconomic or unreliable materials selection. The approach in this document is therefore to identify and control the parameters that are necessary to characterise an intended service environment in terms of its impact on SSC/SCC.

The primary environmental parameters to be defined and controlled are: pH, chloride concentration (Cl^-), temperature and partial pressure of H_2S (p_{H_2S}). Service limits for each material tested can then be presented in the format of Figure 7.1.

Test environments may be application specific or the reference environments proposed in Appendix 4 may be used.

7.3 Test Stresses and Loading of Specimens

The stress or strain applied to a test specimen cannot accurately simulate the mechanical conditions that a component may experience in service. Actual service

pH

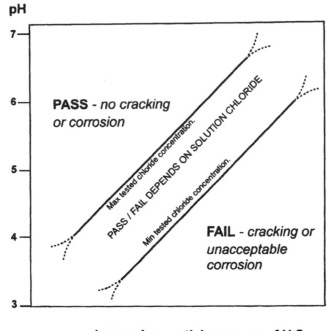

Increasing partial pressure of H₂S
(P_H₂S, arbitrary units)

Fig. 7.1 *Basis of SSC/SCC Tests. Schematic only to show the expected influence of pH, p_{H_2S} and Cl⁻. (Note: The test temperature requires to be stated. The nature and detail of the characteristic will be material dependent).*

stresses are the combined result of a component's manufacturing history and subsequent service exposure.

Service stresses may be:
- static, transient or dynamic;
- applied and/or residual;
- long range or localised.

The maximum stress that coincides with H₂S service may be elastic (with or without prior yield) or plastic (if localised yielding occurs). The extent and occurrence of any plastic straining is important as the cracking resistance of CRAs may be significantly reduced if surface films are mechanically disrupted. Small amounts of plastic straining may occur (as microcreep) under nominally static, elastic loading. Insufficient is known about the mechanical details of service conditions to attempt to specify accurate test simulations that are representative of all applications. In general it would be impractical and uneconomic to try to specify such tests. Clearly, for most components, nominal, static design stresses are unlikely to be an appropriate basis for simple tests as they are not representative of the actual mechanical service condition of the material.

Experience and judgement therefore have to be used to identify appropriate, test conditions for routine qualification of CRAs. To be practical these need to be based on established test methods and facilities.

In assessing test requirements a distinction has to be made between the influence of static stress (or strain) and straining, i.e. a strain rate. In tests, straining may occur in an uncontrolled manner (e.g. through micro creep in a nominally static test), or be controlled (e.g. in an SSRT).

Recognising all these considerations, the primary test methods recommended for the evaluation of the SSC/SCC resistance of CRAs are nominally static tests based on constant load or constant total strain[2]. The recommended test stresses are:

- Constant load tests and sustained load tests: 90% of the actual yield stress of the test material;
- Constant total strain tests: 100% of the actual yield stress of the test material.

Different stresses are recommended in recognition of the known difference in severity of the test methods which largely results from the different amounts of (uncontrolled, plastic) microstraining that can occur by (low temperature) creep relaxation during tests.

This is illustrated for representative, generic materials in Figure 7.2(a, b and c). At stresses near yield, microcreep results from the strain rate dependence of the stress/strain characteristic which is shown in Figure 7.2a(i). Using data for a duplex stainless steel, the extent of microcreep that may occur in constant load and constant total strain tests is shown in Figures 7.2a(ii) and (iii) respectively It can be seen that for stresses near yield, the microcreep that may occur in a constant load test is substantially more than that which may occur in a constant total strain test starting at the same applied stress. The examples shown use ambient temperature data; similar effects will occur in tests at elevated temperatures.

Sustained load tests, in which specimens are loaded by e.g. proof rings, have a characteristic in between the two previous cases. Sustained load tests have been assigned the same loading requirement as constant load tests on the basis that the relative compliance of the (stiff) specimen and (less stiff) ring will make them correspond more closely to the constant load case. The following should be noted with regard to the recommended tests.

i. The tests are not exact equivalents as they involve different stress, strain and straining conditions. The lower stress specified for constant load tests recognises that more low temperature creep occurs in these tests than in constant total strain tests.

---

[2]These are the preferred terms per ISO 7539. See Section 1.2 for definitions.

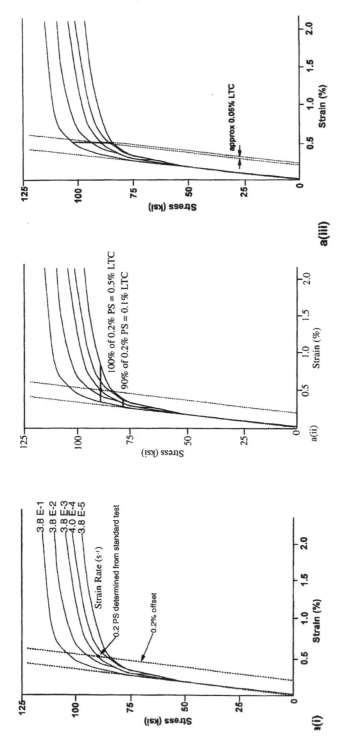

Fig. 7.2a (i) Opposite. Variation in yield strength of a Solution annealed super duplex stainless steel (UNS S32750) at 24°C as a function of strain rate (4.0 × 10⁻³ s⁻¹ approximates to the strain rate typically used in the determination of the 0.2% proof stress). It should be noted that duplex stainless steels exhibit a much greater strain rate dependence of yield strength than other CRAs. The cases described in 7.2a(ii) and 7.2a(iii) represent worst case conditions as they assume instantaneous/rapid loading. Data supplied by Shell Research/MECHANICA.

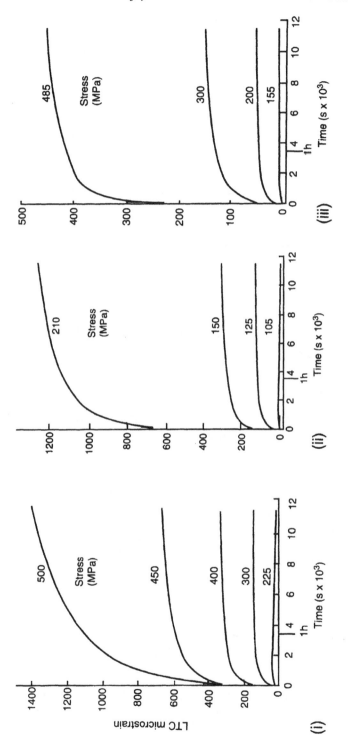

Fig. 7.2b Low temperature creep as a function of stress level at ambient temperature for: (i) 22Cr duplex stainless steel (yield stress = 525 MPa); (ii) 316L austenitic stainless steel (yield stress = 250 MPa); (iii) API 5L X60 pipeline steel (yield stress = 485 MPa). After MM. Festen et al., Environment Induced Cracking of Metals, NACE 10, 1988, p.229.

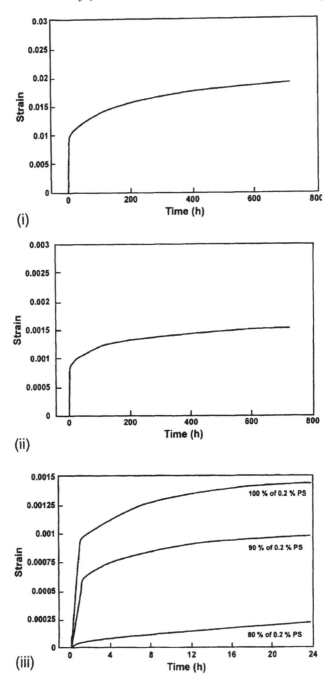

Fig. 7.2c *Low temperature creep at different stress levels at ambient temperature: (i) 22Cr duplex stainless steel (yield stress = 537 MPa) at 0.2% proof stress; (ii) API 13Cr 80 (yield stress 599 MPa); (iii) API 13Cr 80 at different stress levels. Data supplied by Sumitomo.*

ii. Despite their non-equivalence, these tests are recommended as alternatives. It is not required to perform tests under both loading conditions.

iii. These test requirements may not be specifically appropriate for all applications, but they are considered appropriate for most oilfield applications. For specific highly engineered components, a reduced test stress, based on analysis of actual service stresses, may be appropriate. The applied, residual and local stresses must be adequately known. If the material's resistance to cracking is marginal, tests which impose some active straining should be considered.

It is also recommended that the SSC/SCC resistance of candidate alloys be investigated by at least one other test method. In particular:

i.   When no failures occur in nominally static tests, it may be appropriate to obtain information about the mechanical conditions that could cause cracking of the candidate material to identify its limitations in the intended service environment.

ii. The sensitivity of the material to straining should be investigated for applications in which such deformation may occur.

The recommended, supplementary, test is a conventional slow strain rate test (SSRT) as described in ISO 7539-7. This test is not intended to simulate service conditions or to provide stand alone pass/fail data for alloy qualification. Its purpose is to provide additional information about the susceptibility of the alloy to SSC/SCC in the intended service environment. This information can be used to assist the engineering assessment of an alloy's suitability for that service.

By way of exception, it is not considered necessary to perform SSRTs on conventional martensitic steels of the 9 and 13%Cr types as the surface films on these generic alloys are generally not fully passive in H_2S service.

These recommendations are summarised in the table at the top of p.28.

7.4 Test Specimens

The proposed test methods can be used with different specimen geometries as appropriate to the product form to be tested. The following specimens are routinely used for common product forms and may be loaded statically or in SSRT (see ISO 7539-7).

i.   Tubing: C-ring and smooth tensile.

ii. Piping components (parent material): C-ring and smooth tensile.

iii. Piping (girth welds): four point bend and smooth tensile.

		Test Method Increasing Severity of Test Due to *in situ* Straining →			
Applied Test Stress		**Constant Total Strain**	**Constant Load**	**ISSRT* CSSRT***	**SSRT**
Increasing	< 90% YS	o	o	o	-
Severity	90% YS	o	**P**	o	-
of Test	YS	**P**	o	o	-
↓	> YS	o	o	o	**S**

P : Recommended (alternative) primary tests.
S : Recommended supplementary test.
o : Optional, alternative tests.
* : See Section 7.10. *Alternative Test Methods.*
YS : Yield strength (see Section 1.1).

iv. Piping (longitudinal welds): C-ring and four point bend.

v. Vessels (parent materials): smooth tensile.

vi. Vessels (welds): four point bend and smooth tensile.

7.5 Test Duration and Strain Rate

The recommended minimum duration of all constant load and constant total strain tests is 720 h.

The recommended strain rate for use in slow strain rate tests is 1×10^{-6} s^{-1}. This rate is an experimentally acceptable compromise; at higher rates mechanical effects predominate and susceptibility to SSC/SCC may be missed, while at lower rates the time required for each test becomes excessive. Results of tests conducted on specimens of different size or configuration, or at different strain rates, cannot be compared directly.

7.6 Test Temperature

No single test temperature can be specified, as the temperature of greatest SSC/SCC susceptibility varies with alloy type, and possibly its product form and the composition of the test environment. Conventional expectations for the influence of temperature on the principal generic grades of CRAs are summarised in Appendix 4 but these should be used with caution. Careful consideration should be given to establishing the temperature of maximum sensitivity to cracking having regard to the different mechanisms that may affect each material.

7.7 Test Materials

Within a grade of material, variations in mechanical and chemical properties will result in changes in the susceptibility of the material to SSC/SCC. Therefore, within a grade, test material should be selected to represent that which is expected to be most susceptible to SSC/SCC. It is recommended that:

i.   Material with the lowest available pitting resistance equivalent (PRE) from the grade is evaluated.

ii. Material with the highest available strength level of the grade is evaluated.

iii. The metallurgy of the material tested is consistent with the grade (as specified or agreed with the supplier). Considerations include: product form, manufacturing route, phase balance and distribution, grain size, etc.

iv. The consequences of in-service ageing should be considered for alloys that have long term exposure to elevated temperatures (e.g. in high temperature wells).

7.8 Interpretation of Test Results

In constant load and constant total strain tests, any cracking is considered to constitute failure. In slow strain rate tests, no loss of ductility or strength is considered to represent immunity to SSC/SCC in that environment. Other criteria for assessment of SSRT results may be defined as appropriate to the intended use of the material.

   Materials that pass the specified test(s) may be considered fit for service in equal or less severe conditions of exposure. Assuming unchanged mechanical requirements, less severe conditions occur (at a given temperature) when:

i.   the p_{H_2S} and dissolved chloride are lower, and

ii. the pH is higher

than the test values.

   Similarly, materials that fail the specified test(s) shall be considered unfit for service in any other equal, or more severe, conditions of exposure. More severe conditions occur when the criteria stated above are reversed.

7.9 Limitations of the Proposed Test Methods

For most applications, the specified 720 h, constant load or constant total strain. Test is expected to be sufficient to establish adequate resistance to SSC/SCC. Limitations of this approach are:

- the test duration may not be long enough to ensure correct evaluation of borderline cases where there is an extended incubation time to initiate cracking.

- the use of a nominally static test load may not be adequate to qualify severe applications in which the material is exposed to plastic straining in service.

When these limitations are of concern it may be appropriate to use longer exposure periods or a test method that imposes some straining.

Inclusion of the supplementary SSRT is intended to assist in assessing the adequacy of the primary, static test results.

It is recognised that the interpretation of SSRT results is difficult. Conventional pass criteria, based on no loss of ductility, no reduction in maximum load, etc., are generally inappropriate for oilfield use. Test failures may result from cracking that occurs only after extensive plastic deformation that is unrepresentative of service conditions. Modified pass criteria, such as a requirement that the normalised ductility ratios, defined in Appendix 5, exceed (e.g.) 0.5, may be used but no correlations exist to permit the satisfactory definition of such criteria in this document. Engineering judgement is therefore required to assess the adequacy of the primary tests in the light of SSRT data.

In the absence of such judgement, the following limits apply to the interpretation of the proposed tests.

i.   Constant load, sustained load and constant total strain tests establish resistance to SSC/SCC under nominally static conditions.

ii. SSRTs may:
     - demonstrate a material to be immune to SSC/SCC when subject to plastic straining;
     - identify materials that are susceptible to SSC/SCC when subjected to plastic straining;
     - provide a comparative measure of cracking susceptibility based on conventional measures such as ductility or maximum load ratios.

Finally, it should be noted that all the recommended test durations are short relative to service lives. Consequently, long-term effects that influence a material's resistance to cracking will be missed. For example, changes in surface films have been reported during long-term exposure of CRAs to H_2S and/or $S°$. These changes may result in localised corrosion and then SSC/SCC from the micro-environment of a pit or crevice.

Such effects require special consideration as they will not be detected by the tests recommended above.

7.10 Alternative Test Methods

The following alternative test methods maybe considered when there is a requirement to evaluate SSC/SCC resistance more fully under conditions involving plastic straining in service. The test environments proposed in this document are valid for performing any of these alternative tests.

i. The SSRT based method of threshold stress determination proposed in ISO 7539-7.

This method of threshold stress determination requires validation for oilfield environments.

- It has been successfully used for determining SCC thresholds in other environments

- Its suitability for identifying SSC thresholds is unknown.

The lack of development of this methodology for oilfield environments has prevented its inclusion in this document.

ii. Modified test methods, based on the SSRT methodology, which avoid excessive plastic straining of the specimen.

Examples based on SSRTs are the interrupted slow strain rate test (ISSRT) and the cyclic slow strain rate (or ripple) test (CSSRT).

- The ISSRT is a variation on the conventional SSRT in which the test is terminated before the specimen fails. The ISO 7439-7 method of threshold stress determination uses ISSRTs.

- The CSSRT is a further variant of the SSRT, in which a specimen is cycled between predetermined maximum and minimum stresses. These may be selected to be reasonably representative of anticipated service stresses, and thus eliminate the severe plastic deformation imposed by a conventional SSRT.

These methods are still under development for SSC/SCC testing of CRAs, so it has not been possible to propose their standardisation in this document.

iii. Fracture mechanics based test methods

Fracture mechanics specimens (such as double cantilever beam specimens) may be used as the basis for alternative test methodologies. Test method 'D', defined in NACE TM0177, is considered appropriate for use with ferritic and martensitic stainless steels if the test environment is modified to conform with the recommendations of this document. The appropriateness of this test methodology is well less established for other CRAs. Therefore, while fracture mechanics testing is recognised as being important for investigating SSC/SCC phenomena, fundamental problems with defining acceptance criteria for such tests have prevented their inclusion in this document.

7.11 Reproducibility of Tests

The difficulty of specifying environmental cracking tests of satisfactory reproducibility is a recognised problem which may only party be explained by the random nature of crack initiation mechanisms. One view is that no tests should be proposed for 'standardisation' until shown to be satisfactorily reproducible. A contrary view has been taken in the preparation of this document. CRAs are subject to extensive testing by manufacturers and users without any consistency of test procedures. It is considered that the definition of a rational basis for such tests, as defined in this document, represents an improvement of the current ill-defined requirements.

8

SSC/SCC Test Procedures

8.1 General Requirements

All reagents, gases, test vessels and fixtures shall be in accordance with ISO 7539-1 or NACE TM0177 as appropriate. When the CRA under test is resistant to general corrosion, relaxation of solution volumes, relative to specimen surface areas, is permitted.

8.2 Test Environments

The test environments may be specific to the intended application or, for more general purposes. One of the reference environments is defined in Appendix 4. In all cases

- To obtain valid and reliable test data, oxygen must be excluded from the test solution. Oxygen levels should be less than 10 wt. ppb (and preferably lower) in the liquid phase.

- Test temperatures should be maintained within ± 2°C.

8.2.1 Application Specific Environments

Application specific test environments should be prepared to meet the following requirements.

i. The test pH should be less than, or equal to, the lowest expected production pH. The minimum desired pH is achieved through the addition of components present in the actual production environment primarily H_2S, CO_2 and bicarbonate added as $NaHCO_3$.

ii. The partial pressure of H_2S should be greater than, or equal to, the maximum expected production partial pressure.

iii. The chloride concentration should be greater than, or equal to, the maximum value that is expected in the produced water. Chlorides should generally be added as NaCl.

The source, nature and analysis of produced water on which test environments should be based are discussed in Appendix 2.

33

8.2.2 Control and Reporting of Test Conditions

The following environmental test variables shall be controlled and recorded:

Partial pressures of H_2S and CO_2 (p_{H_2S}, p_{CO_2})

Temperature

Test solution pH, the means of acidification and pH control. All pH measurements shall be recorded.

Test solution formulation or analysis.

Elemental sulphur ($S°$) additions.

Galvanic coupling of dissimilar metals. The area ratio and coupled alloy shall be recorded.

In all cases the p_{H_2S}, pH, chloride concentration and $S°$ additions shall be at least as severe as those of the intended application. It may be necessary to use more than one test environment to achieve qualification for a particular service.

The following test environments may be used either to simulate intended service conditions or, using nominated conditions, when intended applications are insufficiently defined to allow their simulation.

8.2.3 Environment 1

Service simulation at actual H_2S and CO_2 partial pressures.

1. Test limits:
   a. Pressure : ambient or greater

2. Test Solution:
   a. Synthetic produced water simulating the chloride and bicarbonate concentrations of the intended service. The inclusion of other ions is optional.

3. Test gas:
   a. H_2S and CO_2 at the same partial pressures as the intended service.

4. pH Control:
   a. The pH is determined by reproduction of the service conditions.
   b. The solution pH shall be determined at ambient temperature and pressure under the test gas or pure CO_2 immediately before and after test.
      i. This is to identify changes in the solution that influence the test pH
      ii. Any pH change detected at ambient temperature and pressure will be indicative of a change at the test temperature and pressure.

8.2.4 Environment 2

Service simulation at ambient pressure with natural buffering agent.

1. Test limits:
   a. Pressure   ambient
   b. Temperature   maximum 60°C
   c. pH   4.5 or greater

2. Test Solution:
   a. Distilled or de-ionised water with sodium bicarbonate ($NaHCO_3$) added to achieve the required pH. Chloride shall be added at the concentration of the intended service.
   b. When necessary, a liquid reflux shall be provided to prevent loss of water from the solution.

3. Test Gas:
   a. H_2S at the partial pressure of the intended service.
      CO_2 as the balance of the test gas.
   b. The test gas shall be continuously bubbled through the test solution.

4. pH control:
   a. The solution pH shall be measured at the start of the test, periodically during the test and at the end of the test.
   b. The pH shall be adjusted as necessary by additions of HCl or NaOH.
   c. The pH shall be maintained within a range of 0.2 pH units.

8.2.5 Environment 3

Service simulation at ambient pressure with acetic buffer.

1. Test limits:
   a. Pressure   : Ambient
   b. Temperature : 24 ± 2°C

2. Test solution:
   a. For general use:
      Distilled or de-ionised water containing 4 g L^{-1} sodium acetate (50 mM NaAc) and chloride at the same concentration as the intended service.

   b. For super-martensitic stainless steels[1] prone to corrosion in solution (a):
      De-ionised water containing 0.4 g L^{-1} sodium acetate (5 mM NaAc) and chloride at the same concentration as the intended service.

---

[1] S. Olsen, J. Enerhaug, Common pitfalls during SSC and pitting testing of supermartensitic stainless steels for use in pipelines, Paper No 02038, *Corrosion 2002*, NACE, 2002.

HCl shall be added to both solutions to achieve the required pH.

3. Test gas:
   a. H_2S at the partial pressure of the intended service.
      CO_2 as the balance of the test gas.
   b. The test gas shall be continuously bubbled through the test solution.

4. pH control:
   a. The solution pH shall be measured at the start of the test, periodically during the test and at the end of the test.
   b. The pH shall be adjusted as necessary by additions of HCl or NaOH.
   c. The pH shall be maintained within a range of 0.2 pH units.

8.2.6 Reference Environments

Environments 1–3 above shall be used when testing is specified in accordance with the reference environments given in Appendix 4
   Recommended reference environments are given in Appendix 4.

8.3 Test Pressures

In determining the partial pressures of test gases, in tests conducted at elevated temperatures, due allowance shall be made for the partial pressure of water vapour.

8.4 Specimen Selection and Surface Preparation and Loading

Parent materials and weldments to be tested should be representative of those intended for use in production and conform to the requirements described in Section 7.7.
   Unless otherwise specified, four point bend and C-ring specimens taken from parent materials shall be evaluated with the 'test' surface in the as-received condition. The only surface preparation recommended is thorough cleaning with a water-based laboratory detergent mix and a non-metallic scrubbing pad followed by degreasing. Specimens taken from welded joints may be examined either with their 'as welded' profile intact (to act as a "natural" stress raiser), or machined flush.
   After final preparation and loading, samples of all stainless steels should be exposed to air for a minimum period of 24 hours prior to exposure to the test solution.[1 and 2]

---

[1]   S. Olsen, J. Enerhaug, 'Common pitfalls during SSC and pitting testing of supermartensitic stainless steels for use in pipelines', Paper No. 02038, *Corrosion 2002*, NACE, 2002.

[2]   J. Enerhaug, S. Olsen, U. Steinsmo, Ø. Grong, A new approach to the evaluation of pitting corrosion of supermartensitic weldments. Paper No. 02039, *Corrosion 2002*, NACE, 2002.

8.5 Loading

All specimens should be loaded slowly to minimise the influence of strain rate on yield properties. The loading requirements below recognise the importance of in situ straining during exposure to the test environment and seek, as far as possible, to control it through standardised procedures.

Yield strengths may be specified in accordance with several different conventions as summarised in Section 1.1. The reference value of yield strength for determination of test stresses is the 0.2% proof stress ($R_{p0.2}$), determined as the 'non proportional elongation' in accordance with ISO 6892.

- The reference value of proof stress shall be the actual value determined on the test material and not the specified minimum or maximum value for the grade of material.

- As cold worked materials may show significant anisotropy, the proof stress should be determined in the direction of the maximum applied test stress.

- The 0.2% proof stress of the parent material should be used as the reference value of yield strength for specimens taken transverse to welds. In the case of dissimilar metal joints, the lower strength, parent material, proof stress shall be used.

8.5.1 Constant Load and Sustained Load Tests

For constant load and sustained load tests the recommended test stress is 90% $R_{p0.2}$.

i. Ambient Temperature Tests
    Constant load specimens shall be loaded after the full test environment has been established. Sustained load specimens shall be exposed to the test environment with the minimum practical delay after loading. As the test result may be influenced by creep occurring after loading, the period between loading and exposure should be the same for any series of tests whose results are to be compared. The period shall always be reported.

ii. Elevated Temperatures Tests
    Specimens should be loaded to the required stress (determined from the properties at the test temperature) at ambient temperature. The test environment (apart from temperature) should then be established at ambient temperature. The temperature of the test cell should then be raised to the test temperature.

8.5.2 Constant Total Strain Tests

For constant total strain tests the recommended test stress is 100% $R_{p0.2}$. Jigs for loading constant total strain specimens shall be stiff with respect to the specimen. Relaxation of the specimen can occur if the loading frame is not sufficiently stiff over the full temperature range of use. Frames can be checked by loading a dummy specimen and remeasuring the deflection after 24 h at the intended test temperature. Loading should be performed as follows:

i. Ambient Temperature Tests
    Specimens should be loaded in accordance with the requirements of Appendix 7. The time between loading and exposure to the environment should be minimised, controlled and reported in accordance with the requirements for constant and sustained load tests.

ii. Elevated Temperatures Tests
    Specimens should be loaded in accordance with the requirements of Appendix 7. The sequence for loading and exposure should be as above for constant load tests at elevated temperatures.

8.6 Primary Test Methods

8.6.1 Method A (Statically Loaded, Smooth, Uniaxial, Tensile Specimens)

Tests using uniaxially loaded, smooth tensile specimens shall be carried out in accordance with the procedure specified in NACE TM0177 (Method A) and the following requirements.

The standard tensile specimens shall be in accordance with those recommended in NACE TM0177 (Method A) except that it is recommended that the shoulder radius should be at least 20 mm to minimise the occurrence of unwanted preferential cracking due to stress concentration at these locations.[3] Specimens of smaller cross sectional area can be used when the product form, test vessel or loading facility prevents manufacture or use of standard specimens.

The source locations of conventional specimens that may be extracted from welded joints are shown in Figure 8.1. For service qualification, it is normally sufficient to test specimens taken transverse to the weld that contain weld metal, HAZ and parent material; (specimen 'T' in Figure 8.1). The other specimens may be used as necessary for more detailed investigations.

8.6.2 Method B (Statically Loaded, 4-Point, Bent-Beam Specimens)

Four-point loaded bent-beam specimens should be prepared as described in ASTM G 39 or ISO 7539-2. The modified double beam configuration, described in ASTM G 39 and shown in Figure 8.2(a), may also be used. (ASTM G 39 cites ASTM STP425, 1967, pp. 319–321 for details). The required deflections for bent-beam specimens shall be established in accordance with Appendix 7. Specimens should be loaded so that the service wetted surface is in tension.

---

[3] Background reasons for the increased radius are given in paper 76 presented at *Corrosion 94*, NACE, 1994.

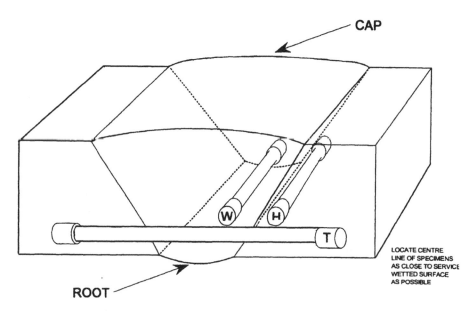

Fig. 8.1 *Designation of tensile specimens taken from welded samples. W: Weld metal; H: Heat affected zone; T: Transverse to weld.*

For specimen T, it is preferred that the weld metal be located in the middle of the gauge length; but, for excessively wide welds, the specimen centre may be displaced to one side of the weld, to ensure that weld metal, heat affected zone (HAZ) and parent material microstructures are all sampled.

Specimen W should contain only weld metal. Specimen H should be taken along the fusion boundary to include the HAZ and some weld metal. Correct location of the specimen should be checked by metallographic examination prior to final specimen preparation. For thin walled pipes it will not be possible to machine tensile W or H specimens and consideration should be given to bent beam specimens.

For welded joints, 4-point loaded bent-beams should normally be taken transverse to the weld, with the weld bead located at the centre of the specimen as shown in Figure 8.2(b) for a conventional 4-point loaded bent-beam. Reduced thickness specimens are permitted, provided that any change in the weld surface condition, resulting from machining is acceptable.

8.6.3 Method C (Statically Loaded, C-Ring Specimens)

C-ring specimens should be prepared as described in NACE TM0177 (Method C), ASTM G 38 or ISO 7539-5. The required deflections for C-ring specimens shall be established in accordance with Appendix 7. Specimens should be loaded such that the service wetted surface is in tension.

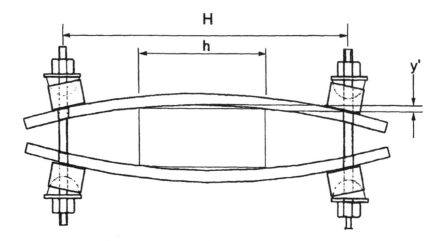

(a) DOUBLE BENT BEAM CONFIGURATION LOADED BY STUDS.
(Alternative to welded configurations shown in ASTM 39 and ISO 7539-2.)

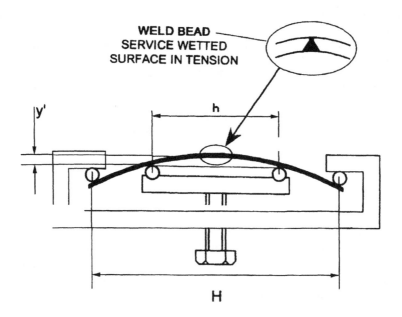

(b) FOUR-POINT BEND SPECIMEN:
STRESS TRANSVERSE TO WELD.

Fig. 8.2 *Schematic illustration of 4-point bend specimens and jigs (based on ISO 7539-2 :1989).*

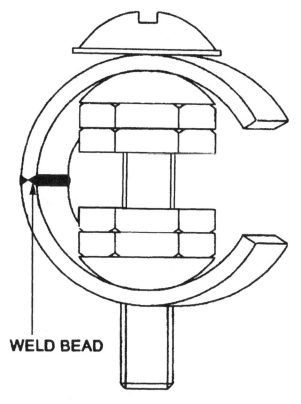

WELD BEAD

Fig. 8.3 Schematic illustration of welded C-ring specimen (based on ISO 7539-5:1989).

For welded joints, C-ring specimens should normally be taken transverse to the weld, with the weld bead located at the centre of the specimen as shown in Figure 8.3.

8.6.4 Examination and Failure Appraisal of Methods A, B and C

The occurrence of any cracking constitutes failure of the specimen.

Specimens that have not clearly cracked through the section shall be examined to establish that the test region is free from any signs of fissures or cracking. All specimens should be examined under a low power microscope (e.g. x10) for indications of pitting and cracking. A sharp probe should be used to uncover pitting in any suspect areas. Freedom from cracking shall also be confirmed by one of the following procedures.

i. Examination in a scanning electron microscope of the surfaces exposed to the maximum tensile stress during testing.

ii. Metallographic examination of the material subjected to the maximum tensile stress during testing by sectioning and polishing. Microsections should be prepared to permit examination of any area where it is not clear whether cracking or corrosion has occurred.

8.7 Supplementary Test Method

8.7.1 Method E (Slow Strain Rate Tests)[4]

In principle, any of the above specimen types can be assessed in a slow strain rate test. The procedure for testing smooth tensile specimens is described below as this is the most common SSRT configuration. The use of other specimen geometries is not precluded.

Smooth specimen, slow strain rate, tensile tests shall be carried out in accordance with the procedures specified in ISO 7539-7 and NACE TM01-98.

Specimens shall be pulled to failure using a slow strain rate tensile test machine. The nominal applied strain rate shall be 1×10^{-6} s^{-1}.

The preferred tensile specimens shall be the 3.81 mm diameter, sub-size specimen defined in NACE TM0177 Method A, except that the shoulder radii shall be at least 20 mm as recommended for full size specimens in Section 8.6.1.

8.7.2 Examination and Failure Appraisal of Method E

Slow strain rate tests shall be assessed by examination of the fracture surfaces and evaluation of the mechanical data.

The fracture faces should be examined in an SEM to determine the fracture mode both at the edges and in the centre. Specimens should also be examined for signs of secondary cracking.

The test data shall also be evaluated by determination of the normalised strain to failure (ε_n) and the normalised reduction in area (RA$_n$). These parameters are defined in Appendix 5.

8.8 Test Report

The test report should contain the following (minimum) information.

---

[4] There is no 'test method D' in this document in order to keep the letter designation of test methods consistent with EFC Publication No. 16.

i.   A description of the test material from which the specimens were taken, including its specification, composition, heat treatment, microstructural condition, product form, manufacturing route and section thickness. If applicable, the welding procedure(s) should also be reported.

ii. The specified mechanical properties and those determined on the test material.

iii. The location and orientation from which test material was extracted from the original stock material, together with the type, size and surface preparation of the test specimens.

iv. Details of the test procedure, method of loading, the applied stress (or strain rate) and the means of its determination.

v.   Full details of the test environment, including:
   • The (liquid phase) water composition (including chloride concentration).
   • The method of deoxygenation and the residual oxygen concentration.
   • The total pressure and the partial pressures of H_2S, CO_2, water vapour and any other gaseous components.
   • The method used to maintain the required gas pressures.
   • The pH at ambient temperature and its method of determination shall be reported.
   • The test temperature.

vi. The sequence used to load the specimens and establish the test environment. The time between specimen loading and exposure to the test environment shall be given.

vii. An appraisal of all exposed specimens with details of the method used to identify cracking of statically loaded specimens (test methods A, B, C) and SSC/SCC of SSRT specimens (test method E).

viii. For SSRTs
   • The strain rate.
   • Representative load/elongation curves for the inert, and test, environments.
   • Normalised ductility parameters (per Appendix 5).

ix. Reference to this and other standards used to establish the test procedure.

APPENDIX 1

Mechanisms of Environmental Cracking

In the presence of H_2S, some CRAs may suffer environmental cracking in oilfield service. This must be strictly prevented. To achieve efficient and cost effective prevention requires recognition that two different mechanisms of cracking can occur. These are called sulphide stress cracking (SSC) and stress corrosion cracking (SCC). They are described here with reference to stainless steels and nickel alloys.

SSC is an extension to CRAs of the well known SSC of carbon steel. It is a form of hydrogen embrittlement, that is, a bulk phenomenon. It is basically a cathodic process, which means that cracking is favoured by an applied cathodic polarisation. The SSC of CRAs is sensitive to the stability of their passive films, and therefore (usually) to the pH and chloride content of the corrosive medium.

SCC is an extension to H_2S service of the well known SCC of CRAs in aerated brines. It is a form of localised corrosion, that is, a surface process. It is basically an anodic process, which means that cracking can be prevented by an applied cathodic polarisation. Like SSC, SCC is sensitive to the stability of the passive film and therefore to the pH and chloride content of the corrosive medium. Additionally, the presence of H_2S may have a significant influence on the threshold conditions for the occurrence of SCC.

For SCC, the mechanism of crack initiation (under marginal conditions) is a mechanically assisted depassivation, due to the natural straining rate induced by the combined action of applied and residual stresses (microcreep under constant load, stress relaxation under constant strain). Cracking may also be initiated by mechanically assisted depassivation resulting from in situ straining.

Normally, the worst case for SSC is around room temperature and the worst case for SCC is at the highest service temperature. However, mechanistic synergies involving interactions between passivity, microcreep, diffusion and hydrogen charging, etc. may induce specific worst cases at intermediate temperatures (e.g. 80-120°C as reported for duplex stainless steels). 'Specific' means that the occurrence of such cases depends on the material, the test environment and the test method, i.e. the applied straining rate and the applied stress).

Ferritic and martensitic microstructures are intrinsically sensitive to SSC, and insensitive to SCC, except under an applied straining rate at, or above, the yield strength. (These microstructures exhibit very little natural microcreep). Conversely, austenitic microstructures are intrinsically insensitive to SSC and sensitive to SCC (as a consequence of extensive natural microcreep). Duplex stainless steels can suffer SSC and SCC; the specific environmental conditions determine which, if either, mechanism occurs. Cold working and precipitation hardening can affect both SSC and SCC.

APPENDIX 2

The Source, Nature and Analysis of Produced Water in Oil and Gas Production

A2.1 Introduction

The approach of this document is to establish corrosion test environments based on the characteristics of the service environments in which it is intended a candidate alloy should be deployed. This appendix is provided as an introductory summary of the characteristics of field environments and the reasons that they differ. Guidance is also given on the determination of the in situ pH which is a key consideration in establishing corrosion test requirements. Finally, as corrosion test environments are routinely derived from conventional water analyses, an outline of their interpretation is provided.

This information is included as, hitherto, it has not been widely appreciated by many who have responsibility for defining corrosion test requirements.

A2.2 The Sources and Characteristics of Produced Waters

Produced fluids are conventionally segregated into hydrocarbon and water phases for analysis. The corrosive significance of CO_2 and H_2S determined in the hydrocarbon phase are well understood. In contrast, the influence of the composition of the water phase on its corrosivity is less well appreciated but nonetheless important to the performance of materials, be they carbon steels or corrosion resistant alloys (CRAs). For this reason, the process of selecting materials must include consideration of the range, or origin, of waters that equipment may be exposed to during service.

A simplified schematic of the sources of oilfield waters is provided in Figure A2.1. The principal sources of water originating as liquid are shown as:

i. Formation waters originating in the hydrocarbon-producing formations and adjacent non-hydrocarbon containing layers.

ii. Injection water supplied from the surface for the purposes of pressure maintenance in oil reservoirs.

These waters will normally contain significant amounts of dissolved solids.

In contrast, water produced, during the early life of gas wells, is normally low in dissolved solids as it does not originate as a liquid downhole, but is predominately formed by condensation from water vapour carried in the gas stream.

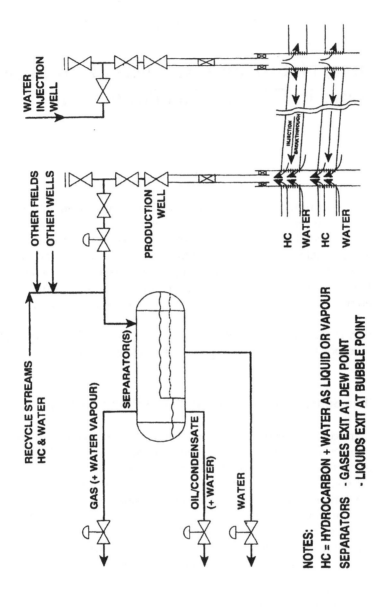

Fig. A2.1 Simplified schematic showing sources of oilfield waters.

A2.2.1 Gas Reservoirs

In gas reservoirs, the hydrocarbons (principally methane and ethane) are in the gas phase and remain in the gas phase in the primary production process. Natural gas liquid (NGL) may be recovered as a separate stream (condensed from the gas) but normally, no stabilised (stock tank) hydrocarbon liquids are obtained.

The hydrocarbon gas is normally/generally water-saturated in the reservoir. The pressure and temperature of the gas fall as it flows to the surface, with the result that liquid water and hydrocarbons are formed by condensation. The condensed water is low in dissolved solids and will achieve low values of pH in the presence of acid gases as it contains little or no dissolved bicarbonate or sulphide to buffer against acidification by CO_2 or H_2S.

Gas wells are generally perforated in the upper part of the hydrocarbon bearing layer(s) to minimise the production of formation water(s). Despite this, dissolved solids (including chlorides) may be present in the water due to entrainment of saline liquid water, originating either in the reservoir formation or adjacent geological layers. In mature fields with active aquifers, water encroachment may change the nature of the produced water. If significant amounts of formation water are produced, the water will have characteristics similar to that produced from oil wells (see below).

As water injection is not economic for pressure maintenance in gas reservoirs, it is not normally a factor to be considered in assessing the produced fluids.

The amount of water produced by gas wells is principally controlled by the water vapour carrying capacity of the gas stream. This is normally enough to cause extensive wetting of the production tubing by a film of condensed water. This film exists in dynamic equilibrium with the gas stream which carries it to the surface as a mist. The composition and pH of the water will be unchanged by contact with CRA tubing; whereas in the presence of carbon steel tubing, a corrosive water will dissolve iron which results in some degree of pH buffering as the water moves up the tubing after its initial formation.

For materials testing purposes, it is common practice to account for small amounts of formation water in the produced water by using test solutions made up from distilled water with an addition of 1 gL^{-1} of NaCl. Although not 'field specific', this generalisation is widely accepted and has been adopted for the 'gas production' reference environment proposed in Appendix 4.

In contrast to the above, it is reported that, in certain circumstances, gas wells may produce waters containing high quantities of dissolved solids under transient conditions. The cause is an accumulation of soluble solids in the liner by precipitation from formation water during normal production. Water ingress may redissolve these solids when the well is shut-in or they may be dissolved in water based fluids introduced from the surface. In either case, water with high dissolved solids, whose concentration can approach saturation, may be present in the well when shut in and/or under restart conditions. Because of this effect, some operators consider that liner and tubing materials must be fully resistant to SSC/SCC in chloride saturated water. The need for requiring CRAs to be tested under such conditions, which may only be transient in service, requires careful justification. Furthermore, it is not clear if, in assessing these conditions, any allowance may be made for any pH buffers (e.g. carbonates) that may be precipitated along with the chlorides.

A2.2.2 Oil Reservoirs

In oil reservoirs, the hydrocarbons are in the liquid phase. The majority of the hydrocarbon mass remains as a liquid in the primary production process which yields a stable liquid under ambient (stock tank) conditions. Associated gas is separated and processed as a secondary stream. A mid-cut of NGL may also be obtained.

In oil wells, water enters the production tubing as a liquid which can originate from any geological formation to which there is communication. Some water will normally be produced from the productive formations. Water may also be produced from non-hydrocarbon bearing formations. Aquifer ingress from below, caused by falling reservoir pressure is a common, and often dominant, water source. The composition of this water may differ from that in the productive layers.

The composition of these waters is specific to the reservoir and may vary significantly from one layer to the next where separate productive layers exist.

In many oilfields, water is injected at a distance from the production wells in order to maintain the reservoir pressure and sweep oil towards the production wells. The composition of the injected water (at surface) is dependent on the field location; the requirements are availability and compatibility with the reservoir rock. Injection water may originate as:

- surface water;
- sub-surface water taken from a water well;
- treated produced water from production wells;
- treated or raw seawater.

Two less common technologies used to enhance oil recovery are:

- Steam floods which involve the injection of water as vapour;
- Miscible floods using 'dry' CO_2 gas which is free of significant water.

These are not considered further in this Appendix.

Generally, it is to be expected that the composition of an injection water will change considerably, through contact with the formation, during its passage to a production well. Although non-native waters will become closer in composition to the formation water, the produced water composition is likely to change significantly when injection water, other than reinjected produced water, breaks through to producing wells.

Compared to gas wells, waters produced from oil wells contain higher quantities of dissolved solids and high chloride levels are common. Additionally, many waters contain bicarbonate which raises the water pH and buffers acidification by CO_2. Water may exceed 90% of the liquid volume produced by an oil well.

There is generally no reliable means of predicting the composition of produced waters originating either as formation water or injected water. For this reason the corrosion engineer must usually depend on samples for information. However, caution is required as samples are notoriously prone to mislead due to either contamination or an inappropriate geological origin. Guidance on sample validation is given at the end of this appendix.

A2.2.3 High Pressure, Gas-Condensate Reservoirs

In high pressure, gas-condensate reservoirs the hydrocarbons are present as a super-critical, 'dense phase' (neither true liquid or gas). Significant amounts of liquid phase 'condensate', NGLs and gas may be produced. Reservoirs may be exploited to maximise either liquid or gas production at surface and the water produced may consequently originate, as described above, for either oil or gas reservoirs. Over time, the water may change from having one characteristic to the other.

Liquids production from reservoirs in which the hydrocarbons have partly segregated by density, is favoured by perforating deep in the productive layer(s). Under these circumstances, the produced water will be predominantly formation water as produced from oil wells[5]. As with oil wells, the volume fraction of water in the produced liquids may be high. As water injection is not used for pressure maintenance in high pressure gas-condensate reservoirs, it does not normally have to be considered.

In contrast to the above, the produced water from these reservoirs may have the characteristics previously described for gas reservoirs. This may result from exploitation of the reserves as a gas resource, usually through perforations placed high in the productive layer(s). Alternatively, as reservoirs initially exploited for liquids production lean out, production is increasingly dominated by gas. Either way, the produced water may be formed primarily by condensation from vapour.

Given these alternatives, it is possible for the composition of water produced by high pressure, gas-condensate reservoirs to change significantly during the life of the field.

A2.2.4 Surface Production Facilities

The nature of surface facilities varies greatly, but normally involves the separation of liquid and gas phases, usually with separate offtakes for water and hydrocarbon liquids as shown in Figure A2.1. Water arriving as liquid at the surface comprises a mix from the various downhole water sources. Its final composition may be modified by such factors as:

i.   Equilibrium with the hydrocarbon phase at surface temperature and pressure.

ii. Mixing of different hydrocarbon and water streams from other wells or fields.

---

[5] It is possible that, during early production, if little or no formation water is produced, a significant part of any water phase detectable at surface may result from the condensation of water vapour from the hydrocarbon phase. The surface water may therefore initially be 'low' in dissolved solids. Although water from this source will always be present, it will not be detected when significant formation water is produced.

iii. Precipitation of 'scaling' ions by mixing of incompatible waters or as a result of changes in temperature and pressure.

iv. Introduction of recycled streams from the downstream process (e.g. 'slops') which may contain water(s) of a different composition. Oxygen and microbiological contaminants may also be present in these streams.

v. Additions and returns of production chemicals (acids, scale inhibitors, deemulsifiers, anti-foams, corrosion inhibitors H_2S scavengers, etc).

vi. Oxygenation by air ingress to low pressure systems.

Although changes of this nature are facility specific and their consequences can not be generalised, there is a systematic difference between the composition of the outlet water in the liquid and gas streams from separators. Whereas the composition of water in the liquid stream will be substantially the same as that of the water phase at the vessel inlet, water in the gas stream is present as vapour which condenses as the stream cools. Thus, in the absence of significant carry over of water as a liquid into the gas stream, the composition of water in the liquid and gas streams will be significantly different. If the inlet water is significantly buffered by dissolved solids, the condensed water in the gas stream may be at a significantly lower pH and, therefore, potentially more corrosive to carbon steels.

A2.3 pH of Produced Waters

Direct measurement of the pH of produced waters at source temperature and pressure is impractical. As recourse the pH may be calculated. (1, 2) or estimated using the graphs given in (3). These are reproduced in Appendix C of EFC Publication No. 16 (4). Equations for estimating the pH of produced waters under CO_2 pressurisation are given in reference (5) for formation water, and reference (6) for condensed water.

A common feature of all these methods of pH determination is a requirement to 'input' the partial pressures of CO_2 and/or H_2S. These should be calculated as:

Partial pressure =   total pressure × mol (or volume) fraction of component in the gas phase.

In the case of liquid streams, at pressures above their bubble point pressure, the gas composition at the bubble point pressure (for the same temperature) should be used as there is no increase in the dissolved concentration of the component at higher pressures.

Two further refinements that may be considered are:

i. In some cases, it may be appropriate to use the gas fugacity, as proposed by reference (6), instead of the (simple) partial pressure.

ii. For CO_2/HCO_3^- equilibria in formation waters, reference (7) provides a correction for the secondary influences of temperature and pressure on pH, for liquids above their bubble point conditions.

It should be noted that all the above methods are based on equilibria determined using CO_2 or H_2S alone. These tend to predict lower (more acidic) pH values than those produced by gas mixes containing substantial amounts of hydrocarbons, as is the norm in oilfield service. Reference (8) indicates that the actual pH of a 0.5 M NaCl solution, under a gas mix containing 23 mol.% CO_2 in methane, will be about 0.2 pH units higher than would be predicted by equilibria determined for pure CO_2. (This is equivalent to halving the effective partial pressure of CO_2). This tendency is generally considered to be an acceptable conservatism when estimating in situ pH values.

A2.4 Application to Material Testing and Selection

The above is a generalised simplification which should be treated as an outline guide only. For individual projects, materials selection should only be made after the anticipated water compositions have been reliably established. Because water compositions are so critical to materials testing, the following section provides guidance on the assessment of water analyses.

A2.5 Water Analysis

A2.5.1 Requirement

By definition, the resistance of materials to aqueous corrosion is dependent on the water composition. The composition of the water phase is therefore a fundamental consideration in assessing the corrosivity of oilfield fluids, and, in defining requirements for corrosion testing of candidate materials for the construction for oilfield equipment. This section considers oilfield water analyses in the context of these requirements.

Chemical analysis of water samples is required to allow interpretation of the in situ water composition. For CRAs there are two primary compositional parameters to be considered for oilfield waters, these are:

i. the chloride concentration, and

ii. the constituents that influence the in situ pH. Conventional water analysis excludes:

- Elemental sulphur which is treated as a separate phase.
- Oxygen and other oxidising species such as Cl and Fe^{3+}, which do not occur naturally in well fluids, but may be introduced as contaminants (with severe corrosion consequences).

- Dissolved acidic gases whose in situ concentrations are inferred from separate analysis of the gas phase.

Water compositions cannot be determined directly under the process conditions encountered in the oilfield. Thus, the in situ composition of water, at service temperature and pressure, has to be evaluated from an analysis made under ambient conditions, on a depressurised sample. The basis for such analyses was established by API RP 45.[9] This document has now been withdrawn as its analytical techniques have been superseded; its general requirements are nonetheless still observed. Further information on analytical methods is available from Ostroff[10] and current ASTM standards.

A2.5.2 Chloride and Non-Scaling Ions

Given a good sample and competent analysis, neither of which should be taken for granted, the analytical values for Na^+, K^+ and Cl^- may normally be taken as reported.

A2.5.3 Scaling Ions

Normally the values for Ba^{2+}, Sr^{2+}, Mg^{2+} and SO_4^{2-} may also be taken as reported as they can be reliably measured and do not directly influence corrosion assessments. The only point to note is that these ions may be 'lost' by precipitation as solids ('scale') when waters from different sources are mixed. This is not of concern when the sampled water has a single source, but should be considered if water compositions, upstream of the point of sampling a mixed stream, are of importance.

Values reported for Ca^{2+} and HCO_3^- require further consideration. The reported values are prone to be 'low' due to precipitation of $CaCO_3$ (carbonate 'scale') at some stage before the analysis is made. Carbonate 'scaling' results from changes of the equilibrium gas pressure and composition, and the water temperature. It is favoured by a reduction of CO_2 partial pressure (increased pH) and increased temperature. Carbonate saturated water will 'scale' in transit from bottom hole conditions to surface sample points, e.g. in flow lines or separators. Further scaling may occur in unpressurised sample containers. Solids are conventionally removed by filtration before analysis of the water, thus Ca^{2+} and HCO_3^- originally dissolved in the water may be lost.

Loss of Ca^{2+} is of little importance as far as corrosion assessments are concerned, but the possibility of a significant associated loss of HCO_3^- should be considered in the context of in situ pH determination as discussed below. The loss of HCO_3^- may be significant when the molar ratio $Ca^{2+}/HCO_3^- \gg 1$, as commonly occurs.

A2.5.4 Iron (Fe^{2+}/Fe^{3+})

Water analyses generally report iron as Fe^{2+}, as that is the form in which iron would originally be present in oxygen-free, produced waters. In practice, Fe^{2+} is normally determined as 'total iron' which includes Fe^{2+} formed by oxidation of original Fe^{2+}.

The reported value will include any contamination (Fe^{2+} or Fe^{3+}) introduced before, or during, sample collection. Contamination in sample containers can be eliminated, but iron may be picked up during flow through downhole tubing or surface facilities. Errors in the reported iron content have little significance for corrosion assessments but oxidation of Fe^{2+} to Fe^{3+} may result in loss of HCO_3^- by unwanted acidisation (see below).

A2.5.5 pH Controlling Components

The main problem with water analyses is accurate determination of the constituents that influence the *in situ* pH of the source water. Conventionally, the pH of water samples is reported. Measurements are made after depressurisation and atmospheric exposure of the sample. These values are useful as an analytical quality control check (see below) but of no use in assessing the *in situ* pH. Generally the *in situ* pH has to be determined by calculations (as previously outlined) which take account of the three controlling buffer systems, these are:

CO_2/HCO_3^-;

H_2S/HS^-;

HAc/Ac^-.

The notations HAc and Ac^- are used for acetic acid and the acetate ion respectively.

It is necessary and convenient to extend the HAc/Ac^- equilibrium to include the other organic acids that may influence the pH determination. These are the volatile fatty acids which include acetic, propionic, butyric and valeric acids in addition to HAc. Since they all have similar dissociation constants, they may be taken as equivalent in the process of pH control. Thus, their molar-equivalent concentrations (in milli-equivalents per litre, $meqL^{-1}$) may be added, and expressed in terms of an acetate equivalent.

To do this it is necessary to convert conventional measurements of concentration given as mgL^{-1} (N), to $meqL^{-1}$ (N') as follows:

$$N'(in\,meqL^{-1}) = \frac{Valency}{Mole} \times N(in\,mgL^{-1}) \tag{1}$$

The acetate equivalent (Ac_e^-) is then determined as the sum of the molar-equivalents, i.e.:

$$Ac_e^- = \sum_{n=2,5} N'\{CH_3(CH_2)_{n-2}COO\}^- \tag{2}$$

More generally, it is recommended that all ionic species are routinely converted to molar-equivalents, (in $meqL^{-1}$ as above), to allow meaningful comparisons of their concentrations.

A2.5.6 Total Alkalinity and Bicarbonate (HCO$_3^-$)

A bicarbonate concentration is routinely reported in water analyses. Historically, the measurement was made by acid titration down to the turning point of methyl orange. A pH meter is now used to determine the end point whose value is typically 4.5. but subject to variation according to the analytical method used.[6] This method produces a value for 'total alkalinity' which is not specifically HCO$_3^-$ and not reliably representative of the in situ HCO$_3^-$ concentration.

If the water does not contain organic acids (see below) and soluble sulphides,[7] the reported value will be a true bicarbonate value (for the sample) in accordance with the titration end point convention.[8]

Soluble sulphides are not routinely determined in oilfield water analyses. If present, they influence both the reported 'total alkalinity' and the in situ pH by their interaction with the bicarbonate ion which may be written:

$$H_2CO_3 \text{ (or } CO_2 + H_2O) + HS^- = HCO_3^- + H_2S \tag{3}$$

In practice, the equilibrium is such that HCO$_3^-$ and HS$^-$ are closely equivalent in their influence on the pH in the presence of CO$_2$ and H$_2$S. Thus, the convention of ignoring soluble sulphide ides is acceptable (in this context) as the HCO$_3^-$/HS$^-$ equilibrium above, is adequately accounted for in the methodology of pH assessment.

For waters that contain significant amounts of organic acids, the reported HCO$_3^-$ measurement requires correction because organic acids contribute to the measured value of 'total alkalinity' as determined by acid titration. Normally, about 2/3 of the organic acid anions (Ac$^-$) are included in the titration result[9] and conventionally reported as HCO$_3^-$. A corrected value for the HCO$_3^-$, content of the sample may be obtained as follows.

i.   The 'organic acid' content of the water must be quantified by analysis. This may be done by liquid chromatography at a pH > 7, by measuring the content of acetate, propionate, butyrate and valerate.

ii. The results should then be expressed as an acetate equivalent (Ac$_e^-$ in meqL^{-1}) as previously described.

iii. The corrected value of HCO$_3^-$ (more accurately 'total alkalinity', as it may include soluble sulphides) is then obtained (in meqL^{-1}) as:

'corrected' HCO$_3^-$ = reported HCO$_3^-$ (i.e. alkalinity down to pH 4.3) – 2/3 Ac$_e^-$ (4)

---

[6] pH 4.5 is specified as the titration end point by API RP 45 [9] and ASIM D 513.

[7] Hydrolysis will result in soluble sulphides being present as hydrosulfides (HS$^-$). They may be detected by the evolution of H$_2$S when acidised.

[8] If the starting pH of the sample is above 8.1, values may be reported for carbonate (CO$_3^-$) and hydroxide (OH$^-$) in accordance with the conventions established by API RP 45 [9], ASTM D 513, etc.

[9] This correction factor, is given in [11]. It is based on an end point at pH 4.3; for an end point at pH 4.5 the factor is reduced by about 0.1.

It should be noted that the correction factor of 2/3 used in this equation is dependent on the end point pH of the acid titration which is taken in this example as 4.3.[10]

A2.5.7 Ionic Strength

When required for in situ pH estimation, the ionic strength of the water is adequately established by conventional water analyses.

A2.5.8 Sampling

The value of a water composition depends as much on a representative sample of water reaching the laboratory, as on the accuracy of the analytical methods. Hence it is important that the sampling procedures must prevent the loss of dissolved components, which may occur by:

- precipitation (e.g. $CaCO_3$, $Fe(OH)_3$);
- acidification by unwanted oxidation[11] which may cause loss of HCO_3^- and HS^-

Precipitation can be prevented by acidising the samples with HNO_3 or HCl at the point of collection. Obviously, when this is done the alkalinity can no longer be measured, so a second (unacidified) sample is required to permit its determination.

Unwanted oxidation can only be prevented by careful anaerobic sampling and transport. Aeration is prevented by using reasonably large glass containers (e.g. 1 L), by filling them through a deeply submerged sample tube to prevent atmospheric exposure, and by over-flushing several volumes before closing the bottle, with a thick, non-porous plug. There must be no league (vapour space) in the closed container. To prevent aeration, the use of plastics bottles and half-filled containers must be strictly prohibited for the collection of water samples.

A2.5.9 Validation and Correction of Analyses

Checks should routinely be performed to validate water analyses. As these can be complex, it is advisable to obtain assistance from a chemist who is experienced in the sampling, analysis and interpretation of oilfield waters.

The first requirement is that the molar-equivalent sum (in meq L^{-1}) of the cations must equal the sum of the anions. An accuracy of 1% is reasonable. A discrepancy of more than 5% indicates problems with the overall procedure for sampling and analysis.

---

[10] For different pH end points (pH_e), the correction factor (F) is given by:

$$F = \frac{\text{Measured alkalinity due to organic acids}}{Ac_e^-} = \frac{1}{1+10\exp(pH_e - pK_{HAc})}$$

where 'exp' denotes an exponent and pK_{HAc}, the dissociation constant of HAc, is about 4.6. For example, if the alkalinity were measured down to pH 3 (instead of pH 4.3), the correction factor would be 0.98 (instead of 2/3), but the 'corrected' HCO_3^- value for the sample would be the same. The key point is that the alkalinity originating from organic acids, must be separated from the HCO_3^-/HS^- alkalinity that controls the in situ pH in the presence of acid gases.

[11] Oxidation of $Fe^{2+'}$ to Fe^{3+}, and subsequent hydrolysis of Fe3+, can acidise the sample between its collection and its analysis, and thus remove a part of the dissolved HCO_3^- and HS^- as volatile CO_2 and

Secondly, the pH of an unacidified, depressurised sample should be between 5 and 9. Above 9, there is a suspicion of contamination by drilling mud or workover fluids. Below 5. there is a suspicion of acidisation. This is a coarse check, but useful nonetheless, for eliminating rogue samples.

Thirdly, an indirect check is to establish that the water analysis is consistent with the reservoir conditions at its source. With a few geological exceptions (e.g. salt domes), *in situ* formation water is normally saturated in $CaCO_3$. Hence, provided Ca^{2+}/HCO_3^- >> 1 so that the reported Ca^{2+} content cannot have been significantly altered by $CaCO_3$ precipitation, the calculated degree of saturation should remain close to 'unity. Much higher values indicate alkaline contamination, and much lower values indicate acidisation. This check can be more discriminating than the previous one based on the reported pH of the water sample.

Finally, for degrees of $CaCO_3$ saturation slightly < 1, some corrections can be attempted. For example, if acidification is due only to hydrolysis of Fe^{3+}, this means that the original HCO_3^- content of the sample was the sum of the reported contents in HCO_3 and Fe^{2+} (in meq L^{-1}). If such a correction brings the degree of saturation under reservoir conditions back to 1, it can be accepted.

When these checks indicate discrepancies, either the water analysis, or the reservoir data, or both, are questionable.

References for Appendix 2

1. J.-L. Crolet and M. R. Bonis, 'pH measurements under high pressures of CO_2 and H_2S', *Mat. Perform.*, 1984, **23**, 35–42.
2. J.-L. Crolet and M. R. Bonis, 'An optimised procedure for corrosion testing under CO_2 and H_2S gas pressure', *Corrosion*, 1990, **46**, 81–86.
3. M. Bonis and J.-L. Crolet, Practical aspects of the influence of in situ pH on H_2S-induced cracking, *Corrosion Science*, 1987, **27**, 1059–1070.
4. *Guidelines on Materials Requirements for Carbon and Low Alloy Steels for H_2S-Containing Environments in Oil and Gas Production*, European Federation of Corrosion publication number 16, The Institute of Materials, 1995; 2nd edition, Maney Publishing, 2002.
5. J. F. Oddo and M. B. Tomson, *JPT*, July 1982, 1583.
6. C. de Waard, U. Lotz and D. E. Milliams, *Corrosion 91*, Paper 577, NACE, 1991.
7. *Applied Water Technology*, First edition, Campbell Petroleum Services, 1986.
8. J. L. Crolet, *JPT*, August 1983, 1553.
9. API RP 45, Recommended Practice for Analysis of Oil-Field Waters. (This document has now been withdrawn).
10. A. G. Ostroff, *Introduction to Oilfield Water Technology*, NACE, 1979.
11. J.L. Crolet, Which CO_2 Corrosion, Hence Which Prediction?, in European Federation of Corrosion Publication No. 13, *Predicting CO_2. Corrosion in the Oil and Gas Industry*, The Institute of Materials, 1994, 9.

APPENDIX 3

Titanium Alloys - Limitations of Use

A3.1 Scope

This appendix summarises specific environments and conditions in which titanium and its alloys should not be used. The information is provided in the context of the general requirements for compatibility of CRAs with produced and non-produced fluids as outlined in Section 6. Compatibility with mercury is not addressed here but recommendations are available in NORSOK standard M-DP- 001 'Design Principles-Materials Selection'.

A3.2 Hydrofluoric Acid and Fluorides

All titanium alloys are rapidly attacked by hydrofluoric acid, even in very dilute concentrations, and also in fluoride containing solutions below pH 7. Titanium should not therefore be specified if exposure to fluorides is anticipated. Gasket and other materials, which may release active fluorides, must not be used with titanium.

A3.3 Methanol

Commercially pure titanium and all titanium alloys are susceptible to stress corrosion cracking in anhydrous methanol containing chlorides or bromides. Cracking by this mechanism is prevented by the presence of a small amount of water in the methanol. For methanol containing traces of chloride, a water content of 2 vol.% will normally prevent cracking of commercially pure grades of titanium. For titanium alloys, a water content of 5 vol.% is considered necessary to prevent cracking. Industrial grades of methanol will normally contain up to 2 vol.% water. Further information has been published[1] and is summarised in the Table A3.1.

A3.4 Hydrogen Uptake

Hydrogen uptake can be particularly detrimental to titanium alloys as it can result in the formation of a brittle hydride phase. In sea water or H_2S service, hydrogen absorption may be caused or increased by:

Table A3.1 *Minimum Water Content Required in Methanol to Prevent Stress Corrosion Cracking of Titanium Alloys.*[1]

Titanium Alloy	Min. Water Content Required (wt%)	
	Intermittent Exposure	Sustained Exposure
Unalloyed (Grades 1,2)	1.5	2.0
Ti-0.3Mo-0.8Ni (Grade 12)	2.0	2.0
Ti-3Al-2.5V (Grade 9)	2.0	2.0
Ti-3Al-2.5V-Ru (Grade 28)	2.5	3.0
Ti-6Al-4V (Grades 5, 23)	3.0	3.0
Ti-6Al-4V-Ru (Grade 29)	5.0	10.0
Ti Beta-C™ (Grade 19)	5.0	10.0

- coupling of titanium to a less corrosion resistant metal;

- cathodic protection systems producing potentials below -0.9V Saturated Calomel Electrode (SCE);

- tensile load or residual stress if hydrogen absorption is occurring;

- pH values less than 3, or more than 12, increase the risk of hydrogen uptake;

- higher temperatures which cause an increase of corrosion at the anode and higher hydrogen activity at the cathode;

- hydrogen sulphide which will accelerate hydrogen uptake in the presence of a cathodic potential.

Reference for Appendix 3

1. W. Schutz, 'Guidelines for Successful Integration of Titanium Alloy Components into Subsea Production Systems', Paper 01003, *Corrosion 2001*, NACE, 2001.

APPENDIX 4

Reference Environments for Comparative (or Ranking) SSC/SCC Testing that is not Application Specific

A4.1 Purpose

The test conditions below are recommended for general use when there is a requirement to undertake corrosion testing in characteristic oilfield environments without reference to any specific application. The environments may be used for evaluation of general and localised corrosion resistance as well as for SSC/SCC testing.

A4.2 Solution Chemistry

The solution chemistry shall be in accordance with section 8.2.2

A4.3 Solution pH

The test solution pH is specified at ambient temperature.

i. The conditions required to establish the specified pH at ambient temperature shall be established by appropriate calculation or direct measurement.

ii. For elevated temperature tests, the solution chemistry and partial pressures of gaseous components shall be the same as those required to achieve the specified pH at ambient temperature. This requires that a correction be made to account for the effect of water vapour pressure at elevated test temperatures.

A4.4 Reference Test Environments

The following are recommended as 'standard' test solutions that are representative of typical service environments for which CRAs may be specified.

i. Typical case for gas production with traces only of formation water.
    pH   :   3.5 (at ambient temperature);
    Chloride   :   1 g L^{-1} NaCl.

Table A4.1

Alloy Type	Dominant Cracking Mechanism/Temperature	Other Considerations
Martensitic Stainless Steels	SSC at ambient temperature.	Cracking reported at higher temperatures for 'super' martensitics[1]
Duplex Stainless Steels	May experience SSC/SCC over a range of temperature. Widely reported to be worst in the range 80-120'C.	SCC at maximum service temperature.
Austenitic Stainless Steels, Nickel Alloys and Titanium Alloys	Main concern is SCC at maximum service temperature.	Titanium Alloys: see Appendix 3.

Note: In all cases the main concern for pitting and crevice corrosion exists at the maximum service temperature.

ii. Typical case for oil production with high dissolved solids from formation water.

    pH   :   4.5 (at ambient temperature);

    Chloride   :   165 g L^{-1} NaCl (100 g L^{-1} dissolved chloride).

The combination of low pH and high chlorides is generally untypical of service conditions but may be useful as a severe, (elemental sulphur free) test environment. For such cases, the following conditions may be considered.

    pH   :   3.5 (at ambient temperature);

    Chloride   :   250 g L^{-1} NaCl (151 g L^{-1} dissolved chloride).

A4.5 Test Temperature

No single test temperature can be specified, as the temperature of greatest SSC / SCC susceptibility varies with alloy type, and possibly its product form and the composition of the environment. Test temperatures should be selected on the basis of the summary at the top of p.63, according to generic alloy type and the property to be investigated, see Table A4.1.

Reference for Appendix 4

1. T. Rogne, et al., 'Intergranular corrosion/cracking of weldable 13% Cr Steel at elevated temperature', Paper No. 02428, *Corrosion 2002*, NACE, 2002.

APPENDIX 5

Normalisation of Slow Strain Rate Test Ductility Measurements

A5.1 Use of Normalised Measurements

In slow strain rate tests, susceptibility to environmental cracking can be assessed by using normalised ductility measurements as defined below. These ratios are only applicable to tests performed on smooth tensile specimens.

A5.2 Normalised Strain to Failure (ε_n)

The ratio is calculated from the load/elongation curves as:

$$\varepsilon_n = \varepsilon_s/\varepsilon_i$$

where ε_s = plastic strain to failure in the test environment and ε_i = plastic strain to failure in an inert environment at the same test temperature.

Note: ε_n = 1 represents a fully resistant material (no susceptibility to SSC/SCC), and ε_n = 0 represents total susceptibility to SSC/SCC.

A5.3 Normalised Reduction in Area (RA_n)

The ratio is calculated from the specimens' dimensions as:

$$RA_n = (RA_s/RA_i)$$

where RA_s = the reduction in area in the test environment and RA_i = the reduction in area in an inert environment.

Note: RA_n = 1 represents a fully resistant material (no susceptibility to SSC/SCC), and RA_n = 0 represents total susceptibility to SSC/SCC.

Determination of this ratio requires accurate measurement of small dimensions as errors will be 'squared' during the calculation of areas. Also, anisotropic deformation may result in non-circular fracture profiles. Therefore:

- At least six measurements of the fracture surface 'diameter' should be made and the average reported. The measurements should be equally spaced around the circumference of the fracture surface.

- The measurements should be made with a vernier travelling microscope or other instrument capable of similar accuracy. Measurements may be made in a scanning electron microscope.

- The method of measurement and the accuracy of the final ratio should be stated. The occurrence of non-circular fractures should also be reported.

APPENDIX 6

Autoclave Testing of CRAs

A6.1 Scope

This appendix provides guidance on autoclave testing which is a routine requirement for evaluating CRAs under oilfield service conditions. The guidance is provided in the absence of a recognised standard practice for performing such tests.

A6.2 Principles

An autoclave is a pressurised test cell, in which specimens can be exposed in controlled environments, at temperatures and pressures above ambient. This capability is required to simulate the service conditions encountered in oil and gas production.

Test materials may be exposed as simple unstressed coupons or as more elaborate stressed specimens for evaluation of environmental cracking resistance. Oilfield test environments usually comprise an aqueous solution in equilibrium with gases of controlled composition.

Conventional autoclave testing exposes specimens to artificial environments for durations which are necessarily short relative to the required service life of (most) oilfield equipment. Care is therefore required in assessing the likely service life of equipment from the results of such tests (see Section 7.9).

As a minimum, autoclaves require facilities for the control and monitoring of temperature and pressure. They may be fully sealed or fitted with various facilities, e.g.

- continuous gas purge or periodic gas replenishment;

- circulation or internal stirring of the test solution;

- electrochemical control or monitoring;

- a 'pull through' capability for live loading of specimens.

Flowing conditions may be simulated in autoclaves by the use of rotating coupons, or, for electrochemical studies, by rotating disk or cylinder electrodes. These are seldom used for CRAs as these materials are relatively insensitive to flow effects in the absence of erosion by suspended solids.

A6.3 Safety

This appendix does NOT address the requirements for safe operation of autoclaves. The hazards involved in operating autoclaves should be carefully and formally evaluated for each facility. Key requirements are:

- Proper design of the equipment including the provision of reliable protection against excessive temperature and pressure and the provision of safe disposal for hazardous gases.

- Definition of the limiting environmental test parameters of the equipment. The limits of temperature, pressure, liquid composition and gas composition should be identified.

- Formal operating and maintenance procedures that recognise the hazards of working with pressurised hot liquids and gases that are toxic and/or flammable. Hydrogen sulphide gas is toxic and flammable; hydrocarbon gases are highly flammable.

- The pressure integrity of pull through autoclaves should be maintained when specimen failure occurs.

A6.4 Test Vessels

The materials used for the construction of autoclaves and their associated equipment must be fully resistant to corrosion by the test environment. Loose liners or internal containers made of suitably resistant materials can be employed inside the primary pressure vessel but consideration must be given to the consequences of wetting of the main vessel through failure of the 'inner vessel', over agitation, condensation, accident, etc.

The internal surfaces of an autoclave and the surfaces of any internal equipment, may absorb or react with H_2S. These large surface areas can significantly reduce the H_2S levels in low H_2S test environments. To prevent this effect interfering with tests, the vessel and its internal equipment may need to be pre-exposed to H_2S. It should also be noted that, in an autoclave that has previously contained H_2S, a test environment which is intended to be 'H_2S free' can become contaminated with desorbed H_2S if the autoclave vessel and equipment are not cleaned thoroughly between tests.

Autoclaves are commonly constructed of UNS N10276 which is normally immune to corrosion in simulated oilfield environments comprising aqueous chloride solutions, pressurised by mixtures of gaseous H_2S, CO_2, hydrocarbons and sometimes nitrogen. Extra precautions are required if elemental sulphur or stimulation acids (HCl, HF) are present in the test medium as they may cause corrosion of UNS N10276.

A6.5 Test Specimens, Loading Grips and Jigs

Stressed specimens must be loaded in such a way that there is no galvanic interaction between them and the loading grips or jigs, the autoclave body, other fixtures in the autoclave, or other specimens. As described in Appendix 7, Section A7.2, this may be achieved by electrical isolation of the specimens or encapsulation of loading devices in a suitable inert material. The methods described in Appendix 7 may be generalised and extended to other specimen types.

Stressing fixtures must be inert to the test environment, so that neither the fixtures, nor the test environment become degraded. Conventionally this requires the use of a matching, or higher alloy, than the test material.

The design of jigs for loading specimens shall take account of the effects of differential thermal expansion between specimens, and jigs made of different material(s).

A6.6 Applied Load-Corrections for Pull-Through Autoclaves

Slow strain rate and other tests performed in pull-through autoclaves, in which the internal pressure significantly exceeds atmospheric, require a correction factor to calculate the true applied stress. The actual stress will exceed that indicated by an external load cell as a consequence of the internal pressure acting on the pull rods. For uniaxially loaded, tension specimens the correction may be made as follows.

σ_s = true stress on the specimen gauge length;
D_p = diameter of pullrod where it penetrates the autoclave;
D_s = diameter of specimen gauge length;
p = autoclave pressure in bar (gauge);
T_L = indicated load in kg from the load cell; and
g = standard acceleration due to gravity, 9.807 ms^{-2}

$$\sigma_s = \frac{4g\,T_L}{\pi D_s^2} + \frac{p(D_p^2 - D_s^2)}{10\,D_s^2}\ N\,mm^{-2}$$

A further correction should be considered to account for frictional forces occurring at the pullrod seal(s). These will tend to reduce the stress applied to the specimen. The magnitude of these forces can be determined by running dummy tests using instrumented specimens.

A6.7 Test Solutions

The most commonly used test solution is sodium chloride dissolved in distilled water. If a fuller simulation of service conditions is required, a synthetic produced water or field water samples may be used. The pH of the test solution is governed by the pressure of the test gas(es) and by any buffering species (such as bicarbonate) present in the solution.

Solutions, based on acetic acid, used for SSC testing at ambient temperature and pressure, should not be used for simulating service conditions above ambient temperature in autoclave tests.

When hydrocarbon liquids are included in the test environment, their contribution to the overall test pressure should be assessed and, if necessary corrections made in a similar way to those required for water vapour. If hydrocarbon liquids are included, the autoclave should be stirred or shaken throughout the test, to mix the liquid hydrocarbon and water phases. Alternatively, specimens may be located in either liquid phase as required, though this requires careful planning as visual confirmation is not possible.

The test solution should be deaerated prior to introduction into the autoclave, transferred carefully to minimise oxygen pick-up and further deaerated in the autoclave. Multiple cycling of vacuum (20 mbar), and oxygen-free (2 vol. ppm) nitrogen purging is one acceptable technique. The requirement for a dissolved oxygen content below 10 wt. ppb is easily achieved by this method.

A6.8 Test Gases

The pressure and composition of test gases have to be controlled. Accurate measurement of the total system pressure is essential. If hydrocarbon gases are used, care should be taken to ensure measurements are not degraded by the formation of hydrates[12] in unheated connections to pressure gauges and transmitters. (Care is also required to ensure that pressure relief devices remain effective).

The required test gas composition may be specified or stated in terms of the required partial pressures of the components. Calculation and control of the partial pressures of gases in the autoclave requires that the water vapour pressure be known. This is negligible below about 60°C, and is given in standard steam tables for pure water at higher temperatures. The presence of significant amounts of dissolved salts in the test solution reduces the water vapour pressure, rendering standard steam tables inadequate. This difficulty may be overcome by direct measurement of the actual water vapour pressure for the specific solution composition and test temperature.

The required gas composition may be obtained by mixing appropriate (pure or mixed) stock gases, or by use of a pre-prepared mix. All source gases must be of high purity.

- At low partial pressures of H_2S, 'pure' gases may be mixed by use of flow meters during gas purging of the autoclave.

---

[12] Hydrates are solids formed by hydrocarbon gases and water. Under pressure, they may form at temperatures above ambient.

- At higher partial pressures of H_2S, gases may be mixed by measurement of the pressure change of the sealed autoclave for each addition. Sufficient time must be allowed for gas absorption into the test solution at each stage.

  The accuracy of mixing by these means will depend on the accuracy of the flow or pressure instrumentation and the 'ideality' of the mix. Use of premixed gases will generally be simpler and more accurate.

The gas mix is used to saturate the test solution. This may be done by one of the following methods.

i. Continuous purging of the mixed gases. This is the best way of ensuring the intended test environment is maintained but requires that the gas outlet be provided with a suitable water vapour condenser to prevent loss of water from the test solution.

ii. Periodic replenishment of the mixed gases. The requirement for an outlet condenser to prevent water loss will be dependent on the replenishment procedure.

iii. Sealing of the charged autoclave after purging with the mixed gases at ambient temperature. After being fully saturated, the autoclave is sealed and heated to the test temperature. Heating will change the partial pressures of the individual gases which should be determined by analysis or calculation in order to establish the required pressure for saturation at ambient temperature.
    - Use of a sealed autoclave, without gas replenishment, is generally acceptable for testing CRAs that consume no significant amount of the active gases.
    - When this method is used it is good practice to analyse the gas to confirm the intended test conditions are achieved. As a minimum, representative test cells should be analysed at the end of the test period.

The test gas composition may be determined by gas phase analysis but care is required in respect of the source pressure, and to prevent inaccuracies occurring due to contamination, water vapour condensation, etc.

The extent to which actual service pressures may be fully simulated by inclusion of 'inactive' gases (typically N_2 or CH_4) in the mix is restricted by the pressure limits of the autoclave and available premixed gases when these are required for initial pressurisation, replenishment or continuous purging. Inclusion of 'inactive' gases will influence the test environment by reducing the solubility of the active gases at a given partial pressure.

A6.9 Test Monitoring

A practical limitation is that no visual observation may be made of the test conditions or specimens in an autoclave. Detailed planning of test procedures is therefore required to achieve the required test conditions. Interventions to inspect specimens, and other interruptions to a test, carry a high risk of oxygen ingress, which can lead to the formation of elemental sulphur in the adsorbed layer of H_2S on specimen surfaces. Test results may then be seriously misleading.

Corrosion of the test specimens can result in a reduction in corrosivity of the test environment due to loss of active components (principally H_2S and CO_2) or pH modification by dissolved corrosion products. Large exposed areas of actively corroding specimens in a small solution volume will aggravate this effect. In practice, when testing CRAs, the main effect is loss of H_2S, due to filming reactions on fixtures and the autoclave, or by precipitation of insoluble corrosion products. It is therefore good practice to monitor the gas and liquid compositions periodically.

Losses of active gases can be corrected by continuous gas purging or periodic replenishment of the gas phase. The test solution may also require replenishment at intervals, especially during long term-tests. As with other interruptions, care must be taken to avoid oxygen contamination during test solution or gas replenishment.

A6.10 Reporting

Details of the test procedure should be fully reported. The minimum requirements are stated in Section 8.7 of the main document.

APPENDIX 7

Stressing of Bent Beam Specimens and C-Rings

A7.1 Scope

This appendix describes procedures for stressing CRA bent beam and C-ring specimens to the requirements of this document.

A7.2 Studding, Nuts and Jigs

All metallic components used for stressing constant total strain specimens, and in contact with them, should preferably be made of the same material as that under test. Where this is not possible they can be made of an appropriate corrosion and cracking resistant alloy provided that there is no galvanic interaction between the different materials or that dissimilar components are reliably:

- insulated from the test specimens, or,
- isolated from the test environment.

Insulation or isolation is required to eliminate unwanted galvanic effects whenever galvanically dissimilar alloys are used for any part of the loading jig.

When studs and nuts are not made of the test alloy, they can be isolated from the environment by full encapsulation in PTFE tape.

For 4-point loaded bent beam tests (Figure 8.2(b)), the specimen may be insulated from the loading frame by use of suitable material (e.g. glass beads, PEEK or ceramic spacers. Quartz has also been used for this purpose).

For double beam tests (Figure 8.2(a)), the central spacer must be made of the same material as the test specimen or a mechanically suitable, electrically non-conducting material (See pp. 40 and 41 for Figures 8.2 and 8.3.).

It should be noted that:

- PTFE is not suitable for spacers above about 80°C, as it is too soft, although it can be used as tape at higher temperatures.

- PEEK can only be used up to about 150°C. Above this temperature, ceramics may be used.

- If heat shrink sleeving is used for the isolation of jigs, it is imperative to select a nitrite-free grade. Many commercial grades contain additions of nitrite to inhibit corrosion of carbon steel. The presence of nitrite may affect the results of stress corrosion cracking tests.

A7.3 Loading: General

Both the ASTM and ISO standards for bent beam specimens and C-rings, (ASTM G 38 and G 39; ISO 7539-2 and 7539-5) give equations relating deflection to stress. However, all these equations assume elastic behaviour up to the 0.2% proof stress. This assumption is not generally valid for CRAs.

CRAs commonly show deviation from elastic behaviour at 50–65% of their proof stress, hence use of the equations in the above standards will result in significant under-stressing. Typically, at a deflection calculated to produce the 0.2% proof stress, only 60–80% of this value will actually be obtained. Welds further invalidate the use of such equations.

It is therefore considered necessary to establish loading requirements for CRAs by the use of strain gauges in accordance with the practice outlined below. There are two partial exceptions to this general requirement:

- Ferritic and martensitic stainless steels generally exhibit elastic behaviour up to near their 0.2% proof stress. The equations from the above standards may be used for stressing homogeneous specimens of these alloys that are free of cold work and welds.

- The NACE equation for loading C-rings at their 0.2% proof stress may be used as detailed below.

In both cases it is recommended that the accuracy of the equations be confirmed by strain gauges applied to representative specimens.

A7.4 Loading: Bent Beams

4-point loaded, bent beam specimens are described in ASTM G 39 (option c) and ISO 7539-2 (option c), and shown in Figure 8.2. There is no generally agreed formula relating deflection to stress, for 4-point loaded bent beams, outside the specimen's elastic range. Nor has an empirical relationship, equivalent to that used for C-rings in NACE TM0177, been validated. Hence, the most reliable method for assessing the stress applied to 4-point loaded, bent beam specimens is to use strain gauges.

The double beam specimen configuration described in ASTM G 39 (option d) and ISO 7539-2 (option d), is a useful, compact variant of the 4-point bent beam. For use in accordance with this document, deflections shall be established by bolting (as shown in Figure 8.2(a)) rather than by welding. Section 10.5 of ASTM G 39 provides reference to details of a bolted assembly.[1]

As for single 4-point bent beams, there is no generally agreed equation for relating the deflection of double beams to stress, outside the specimen's elastic range. Hence, it is again necessary to use strain gauges to determine test deflections.

When using bent beam specimens, it is advisable to avoid excessive bending of the beam. This can be achieved by conforming to the following:

$h \geq 0.5H$ and $y' < 0.1 h$

where h = spacing between inner supports, H = spacing between outer supports, and y' = maximum deflection of the beam between the inner supports. These dimensions are defined in ISO 7539-2 and shown in Figure 8.2.

A7.5 Loading: C-Rings

C-ring specimens are described in ASTM G 38 and ISO 7539-5 and shown in Figure 8.3. Generally loading requirements for CRAs should be determined by strain gauging representative specimens. This method is valid for all stresses and specimen types.

As an alternative, subject to the limitations stated below, the equation in Section 10.5 of NACE TM0177-96 may be used for stressing specimens at their 0.2% proof stress when the material is not elastic up to the proof stress. At the 0.2% proof stress, the deflection of a C-ring is defined as follows:

$$D=\frac{\pi d(d-t)(R_{p0.2}+0.002E)}{4tE}$$

where D = deflection, d = outside diameter, t = wall thickness, $R_{p0.2}$ = 0.2% proof stress, and E = elastic modulus in tension (Young's Modulus).

This equation gives the deflection at the 0.2% proof stress only, for stressing the outer surface of C-rings made from homogenous materials. It cannot be used for applying other values of stress. It is satisfactory to use room temperature properties to set the test strain, as it has been shown that strain at the proof stress is relatively invariant with temperature. There is no generally agreed equation for stressing C-rings between the elastic limit and the proof stress. Strain gauging should therefore be used for such requirements.

A7.6 Welded Specimens

There are no generally accepted equations for the stressing of welded bent beams and C-rings. The equations in the ASTM and ISO standards, and the above NACE equation for C-rings assume homogenous material with uniform mechanical properties. As welded specimens conform to neither of these requirements, the only reliable way of stressing them is to use strain gauges.

When the principal applied stress is transverse to the weld line, the weld must be positioned at the centre of the stressed section of bent beams and C-rings, as shown in Figures 8.2 and 8.3.

A7.7 Strain Gauging

General guidance for strain gauging is given in BS 6888-1990. Strain gauges should be selected to suit the CRA being tested. 'Steel' type have usually been found to be the most appropriate for the majority of CRAs, but care is needed to minimise temperature fluctuations during use. No generally agreed formula exists for converting strain to stress for CRAs which do not follow elastic behaviour up to the proof stress. The strain at the required test stress should therefore be taken from a stress/strain curve for the actual material under test.[13]

Fixing strain gauges usually requires some degree of surface preparation whose influence on the test should be considered. (Guidance on surface preparation required for strain gauges is given in reference (2)). In particular, if it is required that the test surface be unmodified in any way, the location and method of attachment of the gauges should be assessed carefully. Alternative methods of establishing the required test strain may have to be considered in such cases.

For non-welded material, the strain gauge should be placed at the apex of C-rings and bent beams. For welded material, the weld metal is generally stronger than the parent metal, so the parent metal adjacent to the weld bead will reach yield stress before the weld metal. Hence, for all specimen configurations, two strain gauges should normally be used, one placed on either side of the weld bead. The gauges should be placed sufficiently far from the weld toes that the measurements are unaffected by any local stress/strain concentration. The specimen should be strained until one of the gauges indicates a strain equivalent to the required test stress, obtained, as previously described, from the parent metal stress/strain curve.

Strain gauges must be removed and the specimen surfaces carefully cleaned prior to exposure of the specimen to the test environment.

It is required that all strains be determined by the use of crossed or T gauges to determine both longitudinal and transverse strains. The total or principal strain is then given by:

$$E_p = E_l + v E_t$$

where E_p = principal strain (i.e. strain at the required test stress), E_l = longitudinal strain, E_t = transverse strain, and v = Poisson's ratio.

---

[13] To be strictly consistent, the stress/strain curve should be determined from strain gauge measurements obtained by the same procedure as that used to determine the strain applied to test specimens. In practice it is acceptable to use (uniaxial) strain measurements obtained from a conventional extensometer (or equivalent) to determine the required test strain. This simplification results in a marginally greater test strain than that obtained from a stress/strain curve derived from strain gauge measurements of principal strain.

When testing several C-rings or bent beams of the same dimensions, from the same heat, it is permissible to generate a deflection versus strain curve with a strain gauge and stress the other specimens to the same deflection. This simplification is only applicable for identically manufactured and prepared, parent metal, specimens and cannot normally be used for welded specimens. Because of the variable characteristics of welds, each welded specimen should be strain gauged individually. Welds made by an automatic process, that produces a consistent weld profile, are a possible exception to this requirement.

A7.8 Reporting

The method of establishing test deflection shall be reported.

References for Appendix 7

1. ASTM STP425, 1967, 319–321.
2. British Society for Strain Measurements, Code of Practice CP1:1992.

SUPPLEMENTARY APPENDIX S1

Test Methods for the Evaluation of the Corrosion Performance of Steels and Non-Ferrous Alloys in the System: Water- Hydrogen Sulphide Elemental Sulphur

Editorial Note

This Appendix has been prepared, independently of the EFC work group, by a working group in the Corrosion Sub-Committee of the German Institute for Iron and Steel (Verein Deutscher Eisenhüttenleute, VDEh). This is an extended version of VDEh's document SEP 1865 which carries the same title as this Appendix. This version provides more information and explanation than SEP 1865 but has no (intentionally) different requirements. Where testing to SEP 1865 is specified, reference should be made to the latest VDEh edition which is available in German and English. It is reproduced here by agreement with VDEh as it has substantial relevance to the main document although there are some clear differences of philosophy and scope.

- It is included to provide a methodology for extending the scope of the main document to testing CRAs in elemental sulphur.
- The text is as supplied by VDEh; no significant changes have been made for incorporation in this document.
- Ownership of the text remains with VDEh. Any comments on this Appendix received by EFC will be passed to VDEh for consideration in future revisions.
- This approach has been taken to allow progress with publication of both documents without further delay for formal review of the VDEh text by the EFC Work Group, and visa versa.

The VDEh document is applicable to carbon steels as well as CRAs; its scope is therefore wider than that of the main document.

The principal philosophical difference between the main document and this supplementary appendix is that the VDEh approach has been to define standard, reference environments for all testing. The main (EFC) document takes a similar approach for ranking tests that are not application specific (see Appendix 4), but recommends use of application specific test environments (when necessary) for service qualifications. The latter approach is taken to prevent rejection of alloys that succumb to corrosion in excessively severe standard test environments. To be consistent with this, the standard VDEh test environments may be varied as appropriate to suit intended service conditions. In particular pH, Cl⁻, temperature, P_{H_2S}, P_{CO_2} may be considered to be test variables. Test durations may also be varied in accordance with the main document. The specimens and test stresses recommended in the main document should be considered where SCC testing is performed to the requirements of this Appendix.

S1.1 General

Results of laboratory investigations and field experience have shown that elemental sulphur can be very aggressive to metallic materials in the presence of water. This aspect plays an important role in the production of sulphur-bearing hydrocarbons. Selection of materials for use under these conditions is mostly based on results of specific laboratory testing or on negative field experience.

Experience gained with a given set of service conditions cannot always be translated into other service conditions. Furthermore, the results of one laboratory very often are not comparable with results of other laboratories, because the testing and evaluation methods vary from laboratory to laboratory.

The presence of elemental sulphur increases the corrosivity of the environment for pitting corrosion, stress corrosion and particularly weight loss corrosion. The dominating parameters influencing test results [1] are:

- the composition of the test environment (chloride and sulphur content);
- the temperature (sulphur modifications);
- the methodology of testing.

Regarding testing methodology, it is significant whether or not elemental sulphur comes in direct contact with the test material.

Since different laboratories may give different importance to relevant test parameters controversies may arise in the qualification of materials. Therefore, test procedures have been worked out which allow service relevant materials qualification for oil and gas production in the presence of elemental sulphur and wet hydrogen sulphide.[2, 3]

S1.2 Description of the Test Methods

S1.2.1 Environmental Parameters

Three groups of test conditions have been defined, based on the various combinations of possible service conditions with respect to sulphur concentration, temperature and flow. Each group of test conditions includes different test temperatures. Some of the test parameters are kept constant for all three groups, because these parameters

Table SI.1 Summary of Corrosion Test Methods by type of Corrosion

Type of Corrosion	Condition of Liquid Flow	Test group	Test Methods[*]
General Corrosion	Stagnant, Convective	1, 2, 3	Exposure in Test Autoclaves without or with Slight Movement of the Medium
Flow Induced Corrosion[**]	Turbulent	1, 2, 3	Rotating cage
Stress Corrosion	Convective, in some cases also turbulent	1	• Constant deflection (C-rings, 3-point (SCC) and 4 point bent specimens, C-bent specimens) • Constant strain rate (smooth tensile specimens)
	Static or high enough to assure intense contact between specimens and sulphur	2	• Embedding of the specimens in finely grained sulphur (stagnant conditions). • Constant load (smooth tensile specimens). • Constant deflection (C-rings, 3-point specimens and 4-point bent specimens, U-bent specimens). • Constant strain rate (smooth tensile specimens).
	High enough to assure intense contact between specimens and sulphur	3	Turbulent dispersion of the liquid sulphur: tensile specimens); • constant deflection (rotating cage); • constant strain rate (smooth tensile specimens).
[*]For further information see Section S1.2. [**]Includes erosion-corrosion, and flow induced localised corrosion.			

are important for the corrosion system, while the variation of these parameters is less important. These parameters are given as follows:

Cl⁻: The aqueous solution contains 250 g L^{-1} NaCl.[14]

---

[14] 250 gL^{-1} is the intended value. The value given in reference [S3] (25 g L^{-1}) is wrong and superseded by this revision.

H₂S: This constituent should always be added to the test environment, the partial pressure at room temperature being either 2 or 16 bara.

CO₂: In the presence of a large excess of H_2S the addition of CO_2 generally will not influence the corrosion attack significantly. For the sake of completeness, CO_2 can be added according to service conditions.

O₂: Oxygen should be excluded from the test environment.

The three groups of test conditions differ mainly in sulphur concentration and temperature (See Table S1.1).

Group 1 relates to conditions where elemental sulphur is physically dissolved in the medium. Sulphur should be added in a concentration of 1 gL^{-1}. Generally, this amount of sulphur is not completely soluble in the test solution. Therefore provisions have to be taken to prevent contact of sulphur particles or droplets with the test samples. Since under these moderate conditions, no corrosion is anticipated at low temperatures, it is sensible to carry out these tests only at elevated temperatures (e.g. 100, 130, 150, 200 and 230°C).

Group 2 is characterised by the presence of solid elemental sulphur in the aqueous solution. It must be arranged that the specimen is in direct contact with the sulphur. To fulfil this requirement, finely crushed solidified molten sulphur (not sulphur powder) should be added to the stagnant solution so that the specimen is embedded in the sulphur. The test temperatures selected in this group are therefore below the melting point of sulphur:[15] 25, 60 and 110°C.

Group 3 refers to conditions where elemental sulphur forms a separate liquid phase. Hence, the recommended test temperatures are 130, 180, 200 and 230°C. The composition of the test solution is the same as in test group 2. To ensure that the specimen definitely remains in contact with the sulphur, the amount of sulphur added to the environment should be at least 100 g L^{-1}.

It may be necessary to modify these test conditions in specific cases, depending on the intended application for the material. These modifications may include:

- CO_2: the partial pressure should match service conditions;
- buffering: only if the service environment is anticipated to contain buffering systems;
- inhibitors: only if inhibitors are used in service or when the test is performed to evaluate corrosion inhibitors.

S1.2.2 Test Parameters/Type of Corrosion

When selecting the test method, consideration should be given to the various types of corrosion that may be stimulated by elemental sulphur. Consideration should also be given to the flow rate of the environment. Table S1.1 lists the test methods, test environments and flow conditions suitable for evaluating materials with respect to different types of corrosion.

---

[15] The melting point of sulphur is in the range 116-123°C at ambient conditions. It does not change significantly at the specified test pressures.

Testing under stagnant conditions. If no movement, or only convective movement, of the corrosion medium is expected in service conditions, specimens of all suitable sizes can be exposed in test autoclaves. Appropriate provisions have to be taken to ensure that the conditions of groups 1, 2 or 3 are met.

Testing under turbulent conditions. If turbulent flow conditions are anticipated, it is advantageous to make screening tests with a rotating cage.[4-6] In a typical coupon holder up to 12 specimens are fixed between two glass powder reinforced PTFE plates (70 mm in diameter), mounted on the axis of a magnetic stirrer for tests in 2 litre autoclaves. The specimens are 50 × 10 × 2 mm in size. The arrangement of the specimens between the plates is such that they form the outer wall of a cylinder. Pipe segments used as specimens should be installed such that the pipe surface is subjected to axial flow when the cage is rotated. The circumferential velocity of the specimens can be used as a measure of the flow intensity. The high turbulence generated in the gap between the specimens is welcome because turbulence also occurs at uneven surfaces in service and causes flow induced localised corrosion. Additional information may be found in the literature.[4-6]

Constant load tests. Constant load tests in autoclaves can be carried out on round bar tensile specimens, e.g. in a setup according to literature.[7] If turbulence in the environment is needed, it can be produced by means of a magnetic stirrer. The constant load can be applied by means of a lever arm or a spring. Additional details may be found in the literature.[8]

Constant deformation tests.[16] C-ring specimens, U-bend specimens or other types of bend specimens can be used for tests, preferably in 2 litre autoclaves. The arrangement of the specimens should be such that there is no contact among the specimens or between the specimens and the autoclave. If turbulence in the test environment is desired, U-bend specimens can be tested in the autoclave using the rotating cage.[5]

Constant strain rate tests.[17] An experimental setup similar to that described for constant load testing is used. The loading however takes place at a constant strain rate in the range 10^{-7}-10^{-5} s^{-1} (preferably 10^{-6} s^{-1}). Strain rates in this range have generally been found to produce the greatest susceptibility to stress corrosion cracking.[9] If these tests are to be carried out under the environmental conditions of test group 3, sulphur shall be dispersed in the aqueous solution.

S1.2.3 Test Procedures

The conditions must be in accordance with the selected test group. When evaluating materials for weight loss corrosion, erosion corrosion, and/or flow-induced localised corrosion, the test period should be not less than five days. In the case of stress corrosion cracking testing, under conditions of constant load, a test period of 200 h is sufficient for screening purposes. To assure exclusion of oxygen, the test environment in the autoclave shall be purged with purified nitrogen or preferably purified argon.

---

[16] In the main document, these tests are referred to as constant total strain tests in accordance with ISO terminology.

[17] In the main document, these tests are referred to as slow strain rate tests in accordance with ISO terminology.

S1.3 Evaluation of Results

The evaluation of the experimental results should be consistent with recommended practices.[10] Surface related mass loss rates and appearances of localised corrosion should be identified quantitatively as to their type and density. Photographic documentation of characteristic types of corrosion is advised to visualise the intensity of corrosion attack.

References for Supplementary Appendix S1

1. G. Schmitt, Effects of Elemental Sulphur on Corrosion in Sour Gas Systems, *Corrosion*, 1991, **47**, 185–308.
2. Steel-Iron Test Specification of the Institute for Iron and Steel (VDEh/Germany) SEP 1865: Test Methods for the Evaluation of the Corrosion Performance of Steels and Non-Ferrous Alloys in the System: Water - Hydrogen Sulphide - Elemental Sulphur.
3. G. Steinbeck, W. Bruckhoff, M. Köhler, H. Schlerkmann and G. Schmitt, Test Methodology for Elemental Sulphur Resistant Advanced Materials for Oil and Gas Field Equipment, *Corrosion '95*, Paper 47, NACE, 1995.
4. G. Schmitt and W. Bruckhoff, Inhibition of Low and High Alloy Steels in the System Brine/Elemental Sulphur/H_2S, *Corrosion '89*, Paper 620, NACE, 1989.
5. G. Schmitt, W. Bruckhoff, K. Fäßler and G. Blümmel, Flow Loop versus Rotating Probes - Correlations between Experimental Results and Service Applications, *Mat. Perform.*, 1991, **30**(2), 85–90.
6. R. H. Hausler, D. W. Stegmann, C. I. Cruz and D. Tjandroso, 'Laboratory Studies on Flow Induced Localized Corrosion in CO_2/H_2S Environments - II. Parametric Study on the Effects of H_2S Condensate, Metallurgy and Flow Rate', Paper 6, *Corrosion '90*, NACE, 1990.
7. G. Schmitt and S. Savakis, Investigations on Hydrogen Embrittlement of High Strength Low Alloy Steels in High Pressure Hydrogen under Static and Dynamic Loading, *Werkstoffe und Korrosion*, 1991, **42**, 605–619.
8. ISO 7539, part 1, *Corrosion of Metals and Alloys*, Stress Corrosion Testing Procedures; 08.1987.
9. ISO 7539, part 7, *Corrosion of Metals and Alloys*, Stress Corrosion Testing; Slow Strain Rate Testing; 12.1989.
10. DIN 50 905, *Corrosion of Metals*, Corrosion Investigations (in German).

For Product Safety Concerns and Information please contact our EU
representative GPSR@taylorandfrancis.com Taylor & Francis Verlag GmbH,
Kaufingerstraße 24, 80331 München, Germany

Printed and bound by CPI Group (UK) Ltd, Croydon, CR0 4YY
01/05/2025
01858465-0001